Carp in Australia

PH: To A.D. Butcher, who saw the writing on the wall, and Andrea Fletcher/Brumley for pioneering the study of carp in Australia.

KD: To Gimme Walter, who made me the scientist I am today.

Carp in Australia

Paul Humphries and Katherine Doyle

CRC Press
Taylor & Francis Group
Boca Raton London New York

CRC Press is an imprint of the
Taylor & Francis Group, an **informa** business

First published 2026
by CRC Press
2385 NW Executive Center Drive, Suite 320, Boca Raton FL 33431

and by CRC Press
4 Park Square, Milton Park, Abingdon, Oxon, OX14 4RN

CRC Press is an imprint of Taylor & Francis Group, LLC

First published by CSIRO Publishing in Australia, March 26, as ISBN 9781486317738. This print edition is published by arrangement with the Commonwealth Scientific and Industrial Research Organisation (CSIRO), Australia.

Text Copyright Paul Humphries and Katherine Doyle 2026

ISBN: 9781041264804 (hbk)
ISBN: 9781041264798 (pbk)

Typeset in Guyot Text
by Envisage Information Technology

Foreword

This book by Paul and Katie clearly demonstrates the value and purpose of looking back as well as moving forward – history is relevant! Reading the chapters, this elderly retired 'carpie' (who was part of the original Victorian Carp Program) can see all the developments and advances in studying this notable and notorious fish – the carp. It is amazing to see now what we didn't see back then, but our universe was different then. The book points out the complex nature of past and current issues, not only biological and environmental but also the multi-faceted issues of interacting with individual people and broad cultures. Early in our research we learnt directly about this while sampling in the Goulburn River around 1980. Our boat had 'Carp Research' in big letters on the side, which let the public know we were out there investigating the current concerns. As we approached two anglers fishing from the bank we slowed down and gave them a wide berth, only to have them start shaking their fists and yelling 'Leave our carp alone!' What an eye-opener! We'd thought that everyone saw carp as a problem.

In our early days of planning what to investigate, we were overwhelmed by what we didn't know, all the possible directions we could take and all the questions we had no hope of answering. This book has renewed the hope of answers – just look at the genetic techniques now being investigated, as well as the move from a state-based approach to a national one.

This book, crammed with knowledge, is a contrast to my time nearly five decades ago, when it took us months to create a bibliography by hand-searching through libraries with their 3×5" index cards, posting letters to get inter-library loans of actual books, doing endless hours of photocopying and then returning books. To put this search into perspective for 'younger' readers, by the end of the Carp Program in 1982 we had funding for the first PC for the whole of Freshwater Fisheries – an Apple 2e with a whopping 24k (yes, 'k') of RAM and of course, no internet! It is a reminder of our very limited data-handling capacity back then even though we had the latest technology – a programmable calculator.

It is pleasing to see so much information in one book. It means this book has much to offer all those interested in carp, in particular, young scientists. It will open your eyes to see what we didn't see, and pose more questions for the next lot of researchers and managers. So, this elderly scientist from the past highly recommends Paul and Katie's *Carp in Australia*. But I suggest a bit of caution when using the word 'eradication'. This word implies it is achievable but carp have taught us one thing for sure – they are here to stay! Let's hope you have been enticed to fully explore this capstone book and learn more about viewing the Earth through an ecological lens.

**Doug Hume (78 y.o. 'carpie' and retired Certified Fisheries Scientist No. 1863,
American Fisheries Society)**

European Carp

By Philip Hodgins

The way that rabbits bred across the land,
a frantic follow-on to Austin's twelve,
so too these toothless low-life types have spawned
right through the Murray–Darling waterways.
Escaping from a pond in New South Wales
they paralleled the post-war immigrant
in timing, drive and so-called lack of taste
by filling spots the locals didn't want
and making do with anything at first.
Ironically these other Euro-spills,
or waves, turned out to be the only ones
who'd eat a fish like this: all clumsy scales
and muddy flesh impossible with bones.
Apparently they'd poach it up in stock
and some of them were even said to think
it was a kind of aphrodisiac.
Nowadays their children's middle-class cuisine
excludes such items from the peasant past
while carp have grown in number, size and range,
displaced at least a dozen other fish,
and been officially declared a pest.
At times they've been so thick the kids have shot
them in the water with their twenty-twos
and in the Murray swimmers standing still
have felt a barbelled vacuum on their toes.
In Yanko Creek they suckle at the edge.
Lined up like rows of piglets on a sow
they filter all goodness from the sludge
and cloud the water so that plants don't grow.
Of course there's no surprise in any of this.
As soon as those experimental few
were flooded into main connecting streams
the system's damage was as good as done.
There was no way that it could stay the same.
But how on earth did European carp
get into Charlie Hughes' remotest dam?

(Philip Hodgins, 2000, New Selected Poems, *Duffy & Snellgrove)*

Contents

Preface

Katie wrote her PhD in 2012 on the potential for native freshwater fish like Murray cod and golden perch to eat carp as a way of controlling the invasive species. As someone who had worked on Australian freshwater fishes for some time, but who didn't know Katie from a river blackfish, Paul was asked by Katie's university to examine the thesis. It goes without saying that she passed. And that was the last that Paul heard of Katie Doyle for some time.

Years passed and Paul forgot about Katie and her thesis. And then in 2018, a new researcher turned up at Charles Sturt University's Albury campus. Paul was introduced to this scientist, Dr Katie Doyle, but being a little slow on the uptake it took him about a week to put two and two together. When he did, he burst into her room, crying 'It's only just occurred to me. You're *that* Katie Doyle!'

Since then, the two of us have frequently talked about all sorts of fish-related topics and issues, as ichthyologists do, sometimes agreeing and often not. Paul likes to group fish into categories. Katie is convinced that all fish are unique. But when Paul came to write *The Life and Times of the Murray Cod*, being mostly ignorant of the wily arts of enticing fish onto hooks, he asked Katie if she and her partner in crime, Cameron McGregor, would co-author the recreational fishing chapter. It was a successful partnership.

During the writing of that book, Paul hassled Katie to write a book on a topic with which she was intimately familiar: carp in Australia. To Paul's surprise, Katie demurred, claiming that she hadn't the time, although she'd love to when work was quieter. We all know that that never happens. So, after finishing writing the cod book, Paul managed to convince a slightly less overworked and enthusiastic Katie that writing *Carp in Australia* was worth it, pointing out that fame and fortune were a mere 90,000 words, 14 chapters and three years away. How could she refuse? And so this book came to pass.

Everyone has an opinion on the common, European (a misnomer) or (simply) just carp. And most of the time in Australia that opinion is preceded or followed by swearing. Especially if the person is an angler. But it wasn't always that way. Carp was one of the first alien freshwater species brought into Australia by Europeans – as early as the 1850s. There were many introductions of carp after that, mostly by the various Acclimatisation Societies that were popular at the time. It appears that the species established small populations in various waters in south-eastern Australia well before the Boolarra carp 'escaped' and things changed completely, but that carp's presence did not warrant a lot of attention. In fact, goldfish, often called golden carp, concerned the general public a lot more in the first half of the 20th century.

But from the mid-1950s, the 'European carp' began to court controversy. Several people in Victoria proposed farming carp for food and selling it to farmers to stock in dams so they could have a bit of sport *and* have a feed. The Victorian Division of Fisheries and Wildlife, especially Director of Fisheries and Game, A.D. (Alf) Butcher, were vehemently opposed to any such proposals. Butcher's fact-finding trip to the US cemented his concerns about carp, and his testimony at the 1962 State Development Committee Inquiry into the Introduction of European Carp into Victorian Waters was clear: don't let it happen. The Inquiry resolved to declare carp noxious in Victoria and to destroy all existing carp. Although more than 1,000 water bodies were poisoned, including the pond in the Melbourne Botanic Gardens, carp had already been released into lakes in Gippsland and north-western Victoria, from where they spread into the Latrobe and Murray rivers. An exceedingly wet period followed and carp spread rapidly upstream and downstream in the Latrobe Valley and the Murray–Darling Basin.

Since that time, state and Commonwealth governments have thrown millions of dollars at the carp problem. Early and most notable participants were South Australia and Victoria, but later the Commonwealth government, all jurisdictions in the Murray–Darling Basin and, in the mid-1990s, Tasmania got on board. A succession of Cooperative Research Centres, the Bureau of Rural Sciences and the Fisheries Research and Development Corporation, as well as state agencies, picked up the carp gauntlet and an enviable record of research and management followed.

Dozens of scientists and natural resource managers have been involved in trying to solve the 'carp problem' over the last few decades, and well over a hundred scientific papers and reports have been written on the subject. There are too many scientists, managers, community members and First Nations peoples to mention here who have been involved with carp research and management over the years, but all their work was vital to the writing of this book. They have been instrumental in our understanding of what carp does, and what can be done about it. In the writing of this book, we have drawn heavily on their work and their words. We have worked with many of them, and acknowledge their commitment to the carp cause.

In addition to the published literature, of which there is a lot, we have drawn on newspaper and magazine articles, newsletters, the results of fishing competitions and minutes from angling club meetings, commercial fishing data, oral histories, books, unpublished reports, institutional and private images, unpublished correspondence lodged in state archives, and people who were on the ground, as it were, when carp crises arose. Discussions with people provided us with the context for those times.

This book begins with an introduction to carp in Australia to set the scene, but the next three chapters (Chapters 1–3) barely mention Australia at all. Why? Because to understand carp in Australia, it is essential to appreciate the context within which carp exists in the rest of the world. Its evolutionary origins and its association with humans dating back many thousands of years in Asia and Europe, together with its biology (Chapter 6), explain its success under natural and artificial aquatic conditions. Its spread beyond its native distribution as people realised its potential as an aquaculture species, and especially the anthropogenic changes to the landscape that made the environment conducive to the proliferation of carp, echoes what we have seen in Australia. This leads us to the introduction of carp into Australia and its long and relatively benign 'sleep' (Chapter 4) before the Boolarra carp in Gippsland got away from the Victorian authorities (not without a fight!). The story of the Boolarra 'carptastrophe' and its aftermath (Chapters 5 and 7) provide a salient lesson that even the best-intentioned and most tireless intervention cannot control this type of invasive species, this 'powerful invader' as John

Koehn called it. We next describe in detail the environmental impacts of carp in Australia and touch on the economic and social aspects (Chapter 8), before giving the one good-news story in the book: the introduction and eventual eradication of carp from Tasmania (Chapter 9). Chapter 10 details the various ways that scientists, natural resource managers and community groups have attempted to control, manage and eradicate carp. We then include two chapters focusing on what may be seen by some as the positive aspects of carp in Australia: recreational fishing (Chapter 11) and keeping ornamental carp or koi (Chapter 12). Chapter 13 closes the book with consideration of the future of carp in Australia. Is there hope that it can be eradicated from Australia, or is it here to stay?

Acknowledgements

We would like to thank the following for making the writing of this book not only possible, but a joy. For their sage editorial advice and guidance, Mark Hamilton and Eloise Moir-Ford, and for their patient, thoughtful and meticulous copyediting and production, Adrienne de Kretser and Tracey Kudis from CSIRO Publishing. For references, data sources and other carp-related leads and information, Paul Brown, Dave Crook, Kylie Hall, John Harris, Mark Hudson, Brett Ingram, Alison King, John Koehn, Mark Lintermans, Tsuneo Nakajima, Stuart Rowland, John Stewart, Ivor Stuart, Jason Thiem, Nick Whiterod and Brenton Zampatti. For use of images in the book, Tim Berra, Warren Chad, Richard Hoffmann, Doug Hume, Dirk Spennemann, Lynda Tucker and Jonah Yick. For wonderful insights into the Victorian Carp Program, Doug Hume and Sandy Morison. For insights and revelations into the Tasmanian carp saga, John Diggle, Wayne Fulton, Andrew Sanger, Chris Wisniewski and especially Jonah Yick, who went above and beyond to help. Deanna Duffy from SPAN at Charles Sturt University for her generosity, patience and hard work producing the fantastic maps. John Trethewie for the photos of developing carp larvae. The Public Records Office of Victoria for access to the Victorian State Development Inquiry and Fisheries and Wildlife Division correspondence. Tony Minter, Boolarra Historical Society, for information on Boolarra Fish Farms. Russel Wurcherpfennig, for information on his grandfather Franz, the Boolarra Fish Farms and his generosity in showing us around the ponds during the research for this book. Staff at the CSU Library, for copious inter-library loans. The NLA's TROVE, which is a wonderful resource for newspapers, magazines, newsletters and the rest. May it be always funded. Rob Walsh for alerting us to papers on copepod predation on fish larvae. Michael Spear for discussion on sleeper invasive populations. Alison King for advice, experience and help with finding obscure carp references – especially some of her own. Dave Crook for suggestions on who to talk to about carp-related issues in New South Wales. We thank Ian Andrews (President, Australian Koi Association), Heinz Zimmermann (Auction Coordinator, Australian Koi Association), Roy Moylan (President, Koi Society of Australia) and Shona Macskasy (Koi Society of Western Australia) for their friendly advice, guidance and permission to use images related to koi-keeping in Australia. Nicole McCasker for always being a willing and wise sounding-board for ideas as the book progressed. Cameron McGregor for guidance of recreational angling for the species. Keith Bell, for chats about his vast experience with carp, supplying images and for being a great mentor to Katie. Janet Hodgins for her kind permission to use her late husband Philip Hodgins' poem *European carp* at the front of our book. Brian and Martin Pribble and Shasta for permission to use Jim Pribble's image. Dick Brumley for permission to use his late wife Andrea Brumley's image. Doug Hume for taking so much trouble, for providing photos and unpublished or hard-to-find material and for agreeing to write the Foreword. He has been absolutely fantastic. Jarrod Chapman, Doug Hume, Cameron

McGregor, Ivor Stuart and Jonah Yick for reading parts or all of the book – although all of the mistakes and omissions are ours. The Invasive Animals CRC for Katie's PhD scholarship and Lee Baumgartner, for his support in allowing Katie to complete this book.

Our families, friends and colleagues, without whose support this book would never have been written. Paul is especially grateful to Glenys, whose seemingly limitless patience has got him through writing another book. Katie especially thanks Cameron, who has been by her side through thick and thin, and their beloved dog, Ruger, an invaluable fisheries research companion. Katie also expresses deep gratitude to her parents, Lance and Christine, for their unwavering encouragement and support throughout her journey.

Introduction

How like fish we are: ready, nay eager, to seize upon whatever new thing some wind of circumstance shakes down upon the river of time! And how we rue our haste, finding the gilded morsel to contain a hook. (Aldo Leopold, A Sand County Almanac, and Sketches Here and There, *1949).*

THROUGHOUT MUCH OF THE WORLD, THE COMMON CARP, *CYPRINUS CARPIO*, is prized as a staple, hardy, easily cultured, fast-growing, nutritious and tasty food fish (Fig. i.1). It is the second-most widely introduced species of freshwater fish globally, having been deliberately or inadvertently introduced to every continent on the planet, except Antarctica (Fig. i.2) (Berra 2001; Badiou *et al.* 2011). In fact, its popularity goes back a long, long way. Humans dined on its flesh upwards of 50,000 years ago (Richards *et al.* 2015); the Chinese at least partly domesticated the species 8,000 years ago (Nakajima *et al.* 2019); and 2,000 years ago Roman legionnaires filled their stomachs with carp caught in River Danube floodplains. Napoleon supposedly quipped that an army marches on its stomach, recognising the importance of his fighting men being well provisioned. In the case of the Roman army, battling hostile Germans and Celts who were less than pleased with the expansion of the Roman Empire, a locally sourced fishy diet perhaps contributed to their success (Balon 1995a). The Romans may have even taken their beloved carp back to Rome with them when they were forced to retreat.

THE RISE AND RISE OF CARP

Whether or not carp moved westwards with the Romans is speculative, but it eventually made its way into Western Europe and in time to England. It became an ornamental and culinary feature in the ponds of Catholic monasteries throughout Europe and *the* aquaculture species

Fig. i.1. A typical specimen of the scaled common carp. Source: Paul Humphries.

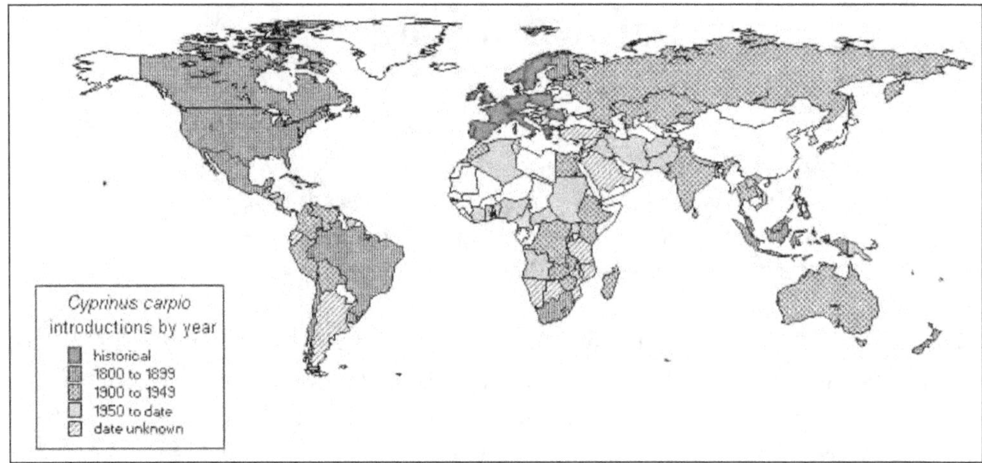

Fig. i.2. Introductions of common carp to countries globally. Note: While the date range may not always be accurate, the map shows the almost worldwide distribution of carp (FAO 2024b).

Fig. i.3. 'Rectangle' the mirror carp, 2008, US. Source: Tim Creque, Ohio, CC BY 2.0 Wikimedia Commons.

in the late Middle Ages (Hoffmann 2023). With a little help from Henry VIII, through his break with the papacy and Dissolution of the monasteries, carp was transformed from a mostly religious fish into one accessible to all. Its success and popularity in England and continental Europe ensured that carp would be a prime candidate for introduction to Australia in the latter half of the 19th century and beyond.

In many places where it has been introduced, the common carp is revered by anglers for its capacity to fight when hooked (Fig. i.3). It was long ago called the 'freshwater fox' because of its wiliness in avoiding the hook. There is today a hugely popular 'coarse' fishery involving carp in many parts of the world, especially in the UK, Europe and North America but increasingly, and perhaps surprisingly, in Australia. Indeed, individual carp have become celebrities in England, anglers obsessively fishing for thousands of hours, hoping to land that particular carp. These icons of the fishy world are mourned when they die, with carp funerals held and eulogies read at their graveside. There are scores of books devoted to angling for carp. Indeed, *CARPology Magazine* recommends '31 carp books you need to own' (https://www.carpology.net/article/other-stuff/31-carp-books-you-need-to-own-/). Talking of magazines, the carpaholic can feed their habit by subscribing to *Total Carp, Big Carp Magazine, Carp-Talk, North American Carp Angler, Crafty Carper, Advanced Carp Fishing* and *Carpworld*. *CARPology* tells us that carp fishing is 'a celebration of the culture of carp angling, offering a non-conformist, self-expressing view of this accessible and credible extreme sport' (https://au.readly.com/products/magazine/carpology). Carp fishing, it seems, is an extreme sport.

A FISH OF MANY COLOURS

The common carp comes in several forms: scaled (fully scaled on the whole body), mirror (patches of large shiny scales), linear (scales only along the lateral line) and leather or naked (scaleless) (Fig. i.4). It has been the subject of artists for centuries (Fig. i.5). While each has a distinctive look, based on its combination of genes, each also has a different growth rate, number of fin rays and even lethal temperature and dissolved oxygen tolerance (Kohlmann 2015).

Fig. i.4. Various forms of the common carp. Source: Unsere Süßwasserfische: eine Übersicht über die heimische Fischfauna nach vorwiegend biologischen und fischereiwirtschaftlichen Gesichtspunkten, by Emil Walter, 1913, Public domain.

The type of carp that gets the heart of non-anglers racing is also one of the most valued and valuable ornamental aquarium fish in the world. Koi carp (a tautology, since *koi* means carp in Japanese), one of the most popular varieties of the common carp, dates back probably only 200 years or so (Balon 1995b) but wow, how much has happened in that short time! The fish's beautiful colours and intricate patterns enchant people in Japan and increasingly in Australia. There is a Children's Day Carp Festival at Tsuetate Onsen, on the island of Kyushu, where carp streamers are strung up over the river. And Japan has a National Koi Day each year on 7 July. The fish is considered good luck and represents prosperity, vitality and longevity. Many countries share the passion for koi, and the finest specimens can fetch eye-watering sums at auction.

Fig. i.5. Mirror with twin carp, about 1200–1225, China, Jin dynasty (1115–1234). Source: Cleveland Museum of Art, CMA 1995.379, CC.

CARP COMES TO AUSTRALIA

Despite carp's popularity overseas, it is almost universally despised in Australia ... with a few exceptions. Apart from the cane toad, it is hard to think of a more hated introduced species. But it wasn't always so. The common carp, and several other species of European fish, was brought to Tasmania and Victoria in the mid-19th century and a little later to New South Wales, by enthusiastic members of acclimatisation societies, whose motives seemed logical and perhaps even conservation-minded at the time (Minard 2013). With the benefit of hindsight, however, the acclimatisers can be thought of as misguided at best, and environmental vandals at worst. But we must be cautious in our criticisms of the actions of people in the past, because there are uncomfortable similarities between more recent controversial conservation ideas, such as 'rewilding' (which may include introducing novel species if they perform a vital ecosystem function), and the motivations of the 19th-century acclimatisers. For decades, the effects of introduced carp were minimal and the fish was considered an exotic oddity, gawked at by an intrigued public in the ponds of botanical gardens or reservoirs and fished for by patient anglers. Most were blissfully unaware of the problems that carp would cause in the future.

Carp remained a relative rarity until the 1960s. Certainly compared to its cousin, the goldfish (*Carassius auratus*) (Fig. i.6), whose at least perceived impact on Australian waterways was much more significant and caused the most angst to the general public in the first half of the 20th century, carp was not on most people's radar for about 100 years after it was introduced. But since the fateful release of the Boolarra strain of the common carp into waterways in Gippsland, Victoria, and into Lake Hawthorn on the Murray River near Mildura in the early 1960s, all hell has broken loose. The fish has spread like a plague throughout much of Gippsland, throughout the Murray–Darling Basin, into coastal catchments in New South Wales and Queensland and even into Western Australia. It has caused, or at least been blamed for, so much environmental destruction that millions of dollars have been spent on attempting to find solutions for this scourge of our waterways. The carp can clearly *drive* environmental degradation through its feeding behaviour, especially in large numbers when confined to shallow lakes and floodplain water bodies (Vilizzi *et al.* 2015). But carp may at times

Fig. i.6. Common carp (top), wild colour goldfish (middle) and wild gold-coloured goldfish (bottom). Note the barbels or whiskers on the common carp and the lack of them on the two goldfish. Note: The carp is >3 kg. Source: Doug Hume.

be a convenient scapegoat – a *passenger* rather than *driver* – for poor river management and environmental degradation more generally, having moved in and proliferated under conditions that are less than optimal for native fish.

With the spread of the Boolarra strain, there has been a tsunami of antipathy in the community toward carp. For those living on the Murray River, their first encounters with masses of carp 'oozing' their way across newly flooded land were disconcerting, worrying and, for many, disgusting (Sinclair 2001).

Riding this wave, some politicians have seen the public relations potential if they can solve the carp problem (or at least be seen to be throwing money at it). Scientists have scrambled to variously eradicate, control, contain or manage the problem. Only in Tasmania, after carp were introduced (again) in the mid-1990s, has eradication been possible (Yick *et al.* 2021). On the mainland, although many imaginative ideas have emerged, including carp separation cages, 'daughterless' gene technology and the cyprinid herpesvirus, no method has so far proved either practicable or palatable to the broader community.

In the wake of the carp wave, some have seen the positive side of a new and abundant fishy resource. Commercial fishing for pet and human consumption and for lobster bait was underway at least as far back as the 1940s, but it boomed with the expansion of the Boolarra carp in the early 1970s. More recently, a few canny entrepreneurs have created lucrative businesses out of this feral species. Keith and Cate Bell, through their company K & C Fisheries Global Pty Ltd, harvest carp commercially for research, supply them to Charlie Carp® fertiliser manufacturers, turn them into animal feed and supplements, and sell fish and fish roe to domestic and international customers. Keith has been in the thick of things since the 1960s when the Boolarra strain of carp was just taking off. The fertiliser Charlie Carp® is well known to gardeners and grew out of an idea by farmers in the Hay and Deniliquin region in 1998. A range of other carp products, including leather and scales, have made their way into national and export markets and are gaining traction.

Common and scientific names of carp

Swedish biologist Carl Linnaeus devised the system of binomial nomenclature in the mid-1700s to avoid the confusion that comes from local or 'common' names for species, so that scientists everywhere would know which species was being referred to. That is why all organisms have a two-part latinised name (the 'scientific name') written in italics. Although people in the past and even today refer to 'carp', that name could mean any number of carp-like species. There is no uncertainty if we refer to *Cyprinus carpio*. Unfortunately, scientists in the past have been guilty of using common names too. *Cyprinus carpio* has variously been called European carp, English carp, Eurasian carp, Prussian carp, mirror carp, koi carp, gold carp, orange carp or just carp. *Carassius auratus* has variously been called goldfish, native carp, gold carp, golden carp, Chinese carp, Gibel carp or just carp. *Carassius carassius* has variously been called crucian carp, Prussian carp, German carp or just carp. *Cyprinus carpio* can be distinguished from the other two species because it has a pair of barbels, or whiskers, on each corner of its mouth. *Carassius auratus* and *Carassius carassius* do not have these barbels, but can be distinguished by the number of gill rakers (finger-like projections on their gills). The former has 19–26, the latter has 14–15. Unfortunately, common carp and goldfish can inter-breed and produce offspring that have a smaller set of barbels on their upper lip.

There are good reasons why people use local or common names. For one thing, the scientific names can be a bit of a mouthful. In this book, we will use mostly just 'carp' as the common name for *Cyprinus carpio*, 'goldfish' for *Carassius auratus* and 'crucian carp' for *Carassius carassius*. For other species, we will make it clear by having the scientific name paired with the common name when first used.

THE CONTRADICTION THAT IS CARP IN AUSTRALIA

With the rise of carp in Australia, a new generation of anglers sees the challenge and fun in catching carp, rather than thinking it just gets in the way of them catching a Murray cod (*Maccullochella peelii*) or golden perch (*Macquaria ambigua*). Some do it as a way of cleaning up the environment while also having the satisfaction of landing a big fish. Others – and an increasingly active cohort – are in it for the sport, emulating their counterparts in Europe, North America and especially the UK. For them, it is certainly about catching big fish, but also about releasing it, allowing it to grow so it can be caught over and over again. Joining an international group of like-minded anglers is part of the attraction. There is the opportunity to compete in international carp fishing competitions held in picturesque places and offering monetary prizes along with the thrill of catching a giant carp.

Finally, there is the burgeoning koi hobby industry that has sprung up in recent years in Australia; a derivative of the enormously popular practice of breeding, keeping and exhibiting *nishikigoi* (brocaded carp) in Japan. The Koi Society of Australia's Annual Koi Show at Fairfield Showground in Sydney attracts thousands of visitors each year. At this and similar shows, you can learn how to keep koi, buy the right equipment to keep fish healthy and exchange advice and ideas with like-minded enthusiasts – while ensuring that none is released into the wild!

You can see and buy koi as cheap as a few dollars or you can spend thousands on one with exactly the right colour and pattern to set off your Zen garden koi pond.

The carp is a contradiction in Australia. Most Australians despise it. Some love it. In many parts of the food bowl that is the Murray–Darling Basin, it is the most abundant fish. Yet most of us turn up our noses at eating it. We have spent millions on trying to control and eradicate it. To date, nothing has worked. Whatever you think of carp, it may well remain a feature of our waterways and increasingly infiltrate our culture for a long time to come, like it has for hundreds of years in other parts of the world where it was not native. So, it's best we learn as much as we can about carp in Australia.

AIMS OF THIS BOOK

Carp in Australia tells the story of how one of the most despised species ever to be introduced to Australia was brought to this country, its relatively recent spread throughout our largest inland river system, the Murray–Darling Basin, its impacts on our river and floodplain ecosystems and the attempts to contain and control it. The book explains why carp has been so popular in Asia and many parts of Europe for millennia and why it has been one of the most widely introduced species of fish globally. It will become clear why carp – among other exotic species – was an ideal candidate for introduction to Australia in the latter half of the 19th century.

Carp in Australia examines the species' origins and tracks its movement into Western Europe from the early Middle Ages in the wake of the rise of Christianity, increasing human populations and the associated agrarian revolution. We show how carp eventually became the go-to fish for aquaculture in continental Europe and England, and was an obvious species for acclimatisation societies in Australia wanting to introduce hardy, fast-growing fish from the 'old country'. We chart the initial relatively benign ripple of the species in Australia, before the tidal wave of the Boolarra strain changed our river systems, perhaps irreversibly.

Carp in Australia includes an outline of the species' biology to provide insight into its success in Australia and elsewhere, and context for the various approaches that scientists and natural resource managers have used, or proposed to use, to deal with this problematic species. We explore the multiple ways that the fish can affect the environments in which it lives, feeds and breeds and look at the evidence on whether it is the *driver* of degradation in the Murray–Darling Basin and elsewhere, a *passenger* that has taken advantage of other environmental stressors, or a bit of both.

Carp in Australia also tells one of the few good-news stories in the carp saga – the introduction to, and eradication of carp from, Tasmania in the very recent past. We describe the attempts by scientists to develop a silver bullet, the role of politics in carp eradication and the somewhat uncomfortable marriage of business and invasive species removal. We illustrate why carp is such a popular angling species and the ornamental lure of koi, especially in Japan and increasingly in Australia. Finally, we look at the future of carp in Australia and see if there is any hope that we might be rid of this most controversial of fish.

1

Origins of the common carp, *Cyprinus carpio*

In [the] twelfth through early fourteenth century carp radiated across most of the economic and cultural heartland of medieval north-western Europe, from the lower Rhine/Mass region south to the Paris basin and Burgundy (Hoffmann 1995, p. 72).

THE COMMON CARP IS NOW ONE OF THE MOST WIDESPREAD FRESHWATER FISH globally (Berra 2001), having been introduced to almost every inhabited continent on the planet (Vilizzi 2012). But its natural wild distribution was much narrower before humans realised what an extraordinary fish it was, and began moving it around. The recorded translocation of common carp began at least 8,000 years ago (Nakajima *et al.* 2019), but likely happened even before then. No doubt its biology – especially its tolerance of poor-quality water and its fast growth – and its tastiness saw it moved beyond its natural range and released into rice paddies or lakes or streams near settlements. At least in some cases, this was done with the very deliberate intent of creating a food source.

We know more about the evolutionary origins and genetics of the common carp (and its family, the Cyprinidae, the most speciose of all freshwater fish families on the planet) than most other species of fish. The common carp is overall one of the most studied freshwater fish, probably rivalled only by the trouts and salmons (Nelson *et al.* 2016). It has been much researched because its cultivation for food and ornamental purposes is important to the economies of many countries (Fig. 1.1) (FAO 2024a). Knowing the breeding habits, growth, genetics and environmental tolerances of the subspecies, strains and variants of common carp is vital to providing the best conditions for fish production (Balon 1995a, b).

Although knowledge from Europe and the West proliferated in English-language journals and books for a considerable time, this has changed dramatically in recent decades with revelations of fish culture practices dating back much further than previously thought (see e.g. Nakajima *et al.* 2019). Where once our understanding of the evolution and spread of the carp relied purely on fossil remains, recent genetic studies and new analytical techniques are giving a clearer picture of the history of this and other fish species (see e.g. Mabuchi *et al.* 2006). New insights will continue to enrich our knowledge of this remarkable fish.

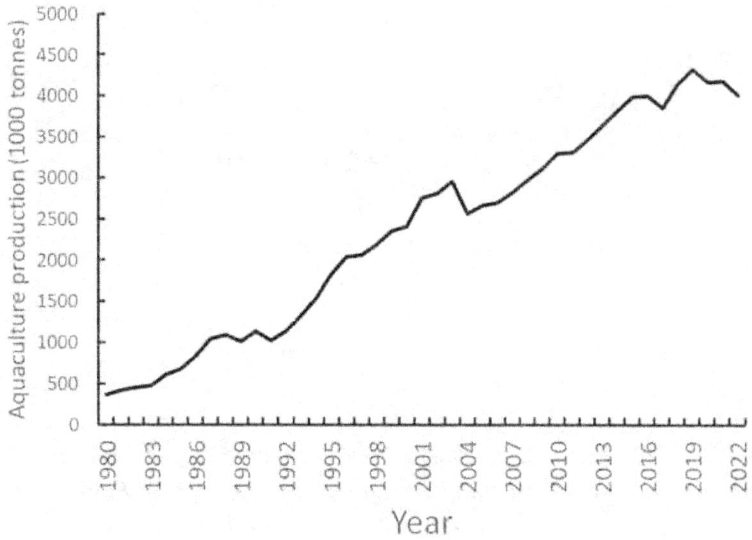

Fig. 1.1. World aquaculture production of common carp, *Cyprinus carpio*. Source: FAO (2024a).

This chapter will describe when and where the common carp originated, to which other group of fish it is related and how it spread – with the help of humans – from its place of origin to cover much of Europe and Asia. In the process, we will learn about how an intersection between the riverine landscape of the mighty River Danube, the biology of a fish and the quest for world domination by the Romans may have kicked off the spread of carp well beyond its natural range. We also will learn how a revolution in agriculture in the early Middle Ages precipitated a wholesale change in the landscape of Western Europe, which facilitated the fish's spread even further. It was ultimately the rise of the carp, especially in Western Europe, that made it a prime contender for introduction to Australia in the 19th century.

EVOLUTION OF CARP AND ITS RELATIVES

The common carp, *Cyprinus carpio*, and its relatives, such as minnows, shiners, chubs, dace, barbs, barbels and numerous other carps, belong to the family Cyprinidae (Order Cypriniformes). Depending on the source, there are as few as two subfamilies (Leuciscinae and Cyprininae) (Cavender and Coburn 1992), or as many as eight (Berra 2001) or 10 (Eschmeyer 2023). The lack of agreement indicates the size of the group and the uncertainty of its origins and relationships. We will try to make sense of things, while bearing in mind that taxonomists periodically change their classification systems and new genetic techniques may emerge.

To date, approximately 1,788 living species (about 3,000 species if extinct ones are included) in 160 genera have been described in the Cyprinidae family (Eschmeyer 2023; Froese and Pauly 2023). According to the fossil record, the family arose during the Eocene, sometime between 56 million years ago and 34 million years ago, most likely in Asia (Zhou 1990), with species appearing in Africa around 40 million years ago (the mid-Eocene) and in North America and

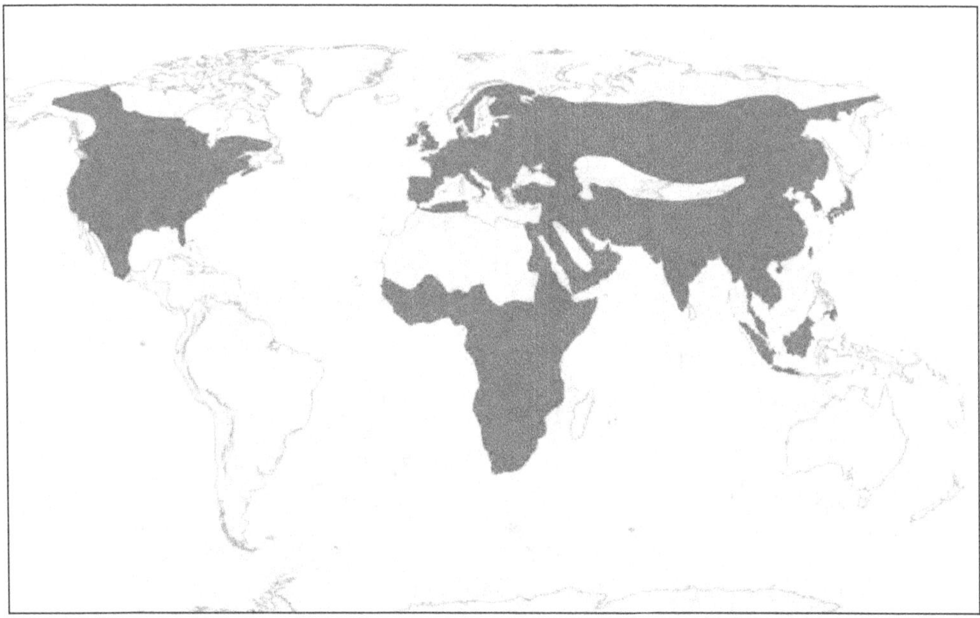

Fig. 1.2. Natural distribution of the world's fishes in the Family Cyprinidae. Source: SPAN, Charles Sturt University, modified after Berra (2001).

Europe around 30 million years ago (the mid-Oligocene) (Cavender 1991). Eurasia has the bulk of the genera and species; approximately equal second are South-East Asia and North America; the fewest genera and species are in Africa (Fig. 1.2) (Howes 1991; Berra 2001). The Cyprinidae's natural distribution, where it typically dominates the freshwater fish fauna, extends throughout Eurasia, Africa and North America (Berra 2001). The family did not occur naturally in South America or Australia, but the introduction of many cyprinid species beyond their natural range has been widespread.

The smallest cyprinid (*Paedocypris progenetica*), which is also a contender for the smallest vertebrate, at maturity barely reaches the size of a thumbnail at 7.9 mm (Kottelat *et al.* 2006), whereas the giant barb (*Catlocarpio siamensis*) may reach 3 m in length and 300 kg in weight (Rainboth 1996). There is an enormous variety of forms of fish in the Cyprinidae, some closely resembling sardines (e.g. *Thrissocypris*), whereas other fish-eating cyprinids look a lot like salmon (e.g. *Elopichthys*) (Fig. 1.3). Cyprinids with which the Australian reader may be familiar include common carp, goldfish, tench (*Tinca tinca*) and roach (*Rutilus rutilus*), all of which were introduced to Australia in the latter half of the 19th century (Harris 2013).

The common carp and goldfish split from other carps (grass carp, *Ctenopharyngodon idella*; bighead carp, *Aristichthys nobilis*; and silver carp, *Hypophthalmichthys molitrix*) sometime between 20 million years ago and 5.6 million years ago (Chistiakov and Voronova 2009; Wang *et al.* 2012; Kohlmann 2015). At the time of the split, common carp and goldfish shared a common ancestor that was tetraploid, which means it had four sets of chromosomes rather than the two that most other fishes and indeed most other vertebrates have. It is thought that tetraploidy came about through hybridisation of two species of the common ancestor of

Fig. 1.3. Various forms of fishes from the Family Cyprinidae.

common carp and goldfish. Common carp and goldfish split some time after that. Because they both are tetraploid, they can inter-breed and hybridise and produce young, which in some cases can themselves reproduce. Common carp (and goldfish) cannot hybridise with other carp species, which are naturally diploid.

WHERE THE WILD CARP ARE

Based on early fossil evidence, the common carp was thought to have originated in the area of the Caspian and Aral seas in Central Asia sometime between 2.6–3.6 million years ago (Fig. 1.4) (Balon 1995b). According to this hypothesis, the species expanded so that its distribution extended west to the piedmont region of the River Danube and to the Amur River and China in East Asia. Later sequences of Pleistocene glaciations separated the western from the eastern populations and they became geographically, and so reproductively, isolated (Berg 1965). This resulted in the subspecies that exist today (Yuan *et al.* 2018). More recent mitochondrial DNA work on a Japanese population in Lake Biwa (see below) has instead supported Japanese ichthyologist Tsuneo Nakajima's hypothesis, based on fossil evidence (Nakajima 1994), for an East Asian origin, splitting from all other Eurasian common carp sometime between 1.7–2.5 million years ago (Mabuchi *et al.* 2005).

Whatever the origins, taxonomists generally agree that there are two subspecies of common carp: *Cyprinus carpio carpio*, the European form, and *Cyprinus carpio haematopterus*, the East

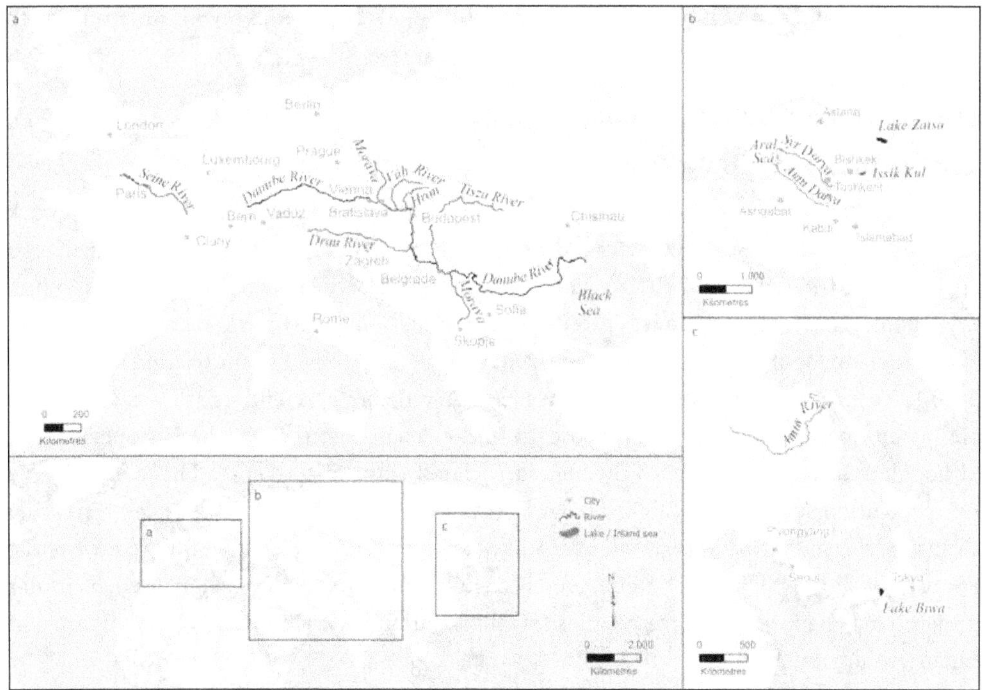

Fig. 1.4. Locations and regions relevant to the evolution and spread of common carp. Source: SPAN, Charles Sturt University.

Asian form. Other subspecies have been proposed (e.g. *C. c. rubrofuscus*, a divergent from *C. c. haematopterus* in China; *C. c. varidivlaceus*, a divergent of the same subspecies in South-East Asia), but not all scientists concur (Chistiakov and Voronova 2009; Kohlmann and Kersten 2013; Kohlmann 2015). Based on genetic analyses, the two subspecies probably separated sometime during the middle or lower Pleistocene, 500,000–900,000 years ago (Kohlmann *et al.* 2003; Zhou *et al.* 2004).

After geographic isolation, the wild populations of the two subspecies spread further westwards and eastwards. The natural distribution of wild populations of *C. c. carpio* included eastern Europe, Turkey, some fluvial basins of the Transcaucasus, the Aral Sea, Syr-Darya and Amur-Darya rivers, lakes Zaisa and Issik Kul, and some lakes in Central Asia and Kazakhstan. The natural distribution of wild populations of *C. c. haematopterus* included the Amur River, many rivers and lakes in China and Vietnam and adjacent countries in South-East Asia (Kohlmann 2015).

There is a lot more known about the spread of carp west of its natural range, largely because of painstaking research by the late Czech-Canadian ichthyologist Eugene Balon, and documents written by various people through the ages. Some of the presumed movement was originally speculative, but has been mostly borne out by subsequent archaeological research. Until recently, less was known of the spread of carp throughout Asia than Europe, although this is changing rapidly with the use of genetics to test hypotheses that were previously just based on fossils (Kohlmann 2015). This catch-up reflects the past dominance of European studies in the broadly available and English-language literature, and the genetics revolution from which we are all benefiting.

ORIGINS AND SPREAD OF CARP IN CHINA AND ASIA
Carp genetics and Lake Biwa

After the split of the western and eastern forms of carp, there was an apparent bottleneck in populations of European wild carp sometime between 100,000 and 10,000 years ago (Kohlmann 2015; Yuan *et al.* 2018). The bottleneck in Asian wild carp, on the other hand, is thought to have been much less pronounced. Populations likely peaked at the start of the last glaciation (about 100,000 years ago) and experienced only a slight decline until about 30,000 years ago (Yuan *et al.* 2018). As we have stated above, scientists are less sure of the timing and patterns of spread of wild carp in the Asian east than in the European west, although to be fair, quite a lot of the ideas associated with western spread have been heavy on speculation and relatively light on data. However, it is probable that, like the retreating Romans who may have spread wild carp into parts of Western Europe to ensure a ready source of fish to which they had become accustomed (Balon 1995a) (see below), a similar scenario may have occurred for peoples travelling throughout China and into Vietnam and Japan (Kohlmann 2015).

Recent genetic studies of carp confirm what had been suspected based on fossil evidence: that the European strains likely have a single origin and the source was Asian (Mabuchi *et al.*

Fig. 1.5. Lake Biwa, Japan, from Mount Shizu. Source: Alpsdake, CC BY-SA 4.0.

Carp scale cover and colour

Genetics is useful in distinguishing species and sub-species, and in carp there are genes that also govern scale cover and colour. Work on common carp genetics has been underway for almost a century, much of it carried out by Russian scientists in the 1920s–1940s. Valentin Kirpitchnikov was at the forefront of research. Carp with complete scale cover, which is normal for wild carp, is 'dominant' in a genetic sense over 'mirror' carp, a form which can be found in domesticated populations. If two scaled carp mate and only have genes for complete scaling, they will always produce a scaled carp. If a scaled carp and a mirror carp mate, up to half of all their offspring may be mirror forms, depending on the genetic make-up of the parents. There are four scale phenotypes (the way an organism looks, rather than what its genetic make-up is): fully scaled carp; 'mirror' or 'mirror scattered' carp; 'linear' or 'mirror linear' carp; and 'leather', 'naked' or 'nude' carp (Kirpitchnikov 1999). The type of scale cover depends on the presence and arrangement of S and N autosomal loci (locations on chromosomes, which are not related to sex but are inherited). Some combinations are lethal. The S and N genes also confer other effects on the various carp forms. Other genes can affect skin colour phenotypes, which produce German blue, Polish blue and red or gold forms – and, the most famous of all – koi (see Chapter 12).

Box Fig. 1.1. Common carp scale cover phenotypes. (A) scaled, (B) mirror, (C) linear or scattered, (D) leather or naked. Source: Casas *et al.* (2013), CC BY-SA 4.0.

2006). However, that is not the case for Chinese, Vietnamese and Indonesian strains. Genetics indicates that there have been 'frequent and wide-scale translocations' (Mabuchi *et al.* 2006, p. 75) of carp into and among these countries and elsewhere. Evidence suggests that the carp in Indonesia are likely to be domesticated strains originally introduced from China or Vietnam in the 17th century (Steffens 2008). The carp in Lake Biwa, Japan, is, however, particularly significant (Fig. 1.5). Tsuneo Nakajima and Kohji Mabuchi have spent decades researching the fish fauna of Lake Biwa, finding that the carp in Lake Biwa forms an 'ancient lineage' – a strain of carp that has existed for a very long time, likely splitting from all other Eurasian common carp 1.7–2.5 million years ago (Mabuchi *et al.* 2005). This is considerably older than the oldest fossil so far found in sediments from a proto-Lake Biwa from 500,000 years ago (Mabuchi *et al.* 2005; Kohlmann 2015). Although more work needs to be done, the Lake Biwa results provide support for East Asia being the place of origin of present-day wild common carp (Mabuchi *et al.* 2005, 2006).

(Really) early domestication of carp?

There is growing evidence of carp aquaculture dating back 8,000 years (Nakajima *et al.* 2019). Carp have been exploited by humans since at least the Upper Palaeolithic (50,000–12,000 years ago) (Richards *et al.* 2015), and raised in ponds and paddy fields in East Asia since around 2000 BCE (Zhao *et al.* 2015). The earliest treatise on fish culture in the world, *Fan li on Pisciculture,* was supposedly written sometime during the 2nd century BCE, combining knowledge accumulated since the 5th century BCE (You 2004, cited in Zhao *et al.* 2015). Nakajima and colleagues speculated that, since rice paddies have existed since 5000 BCE, it may be that carp were actually kept in rice paddies from that time. Evidence from Neolithic sites in China dating from 6000–2000 BCE indicates that the most common fish harvested were at sexual maturity. But other sites, such as the one at Asahi, Japan, dating from around 400 BCE–100 CE, indicate that people were harvesting fish at two main size classes. Nakajima *et al.* (2019) summarise their thoughts thus:

> *In such fisheries, a large number of cyprinids were caught during the spawning season and processed as preserved food. At the same time, some carp were kept alive and released into confined, human-regulated waters where they spawned naturally and their offspring grew by feeding on available resources. In autumn, water was drained from the ponds and the fish harvested, with BL [body length] distributions showing two peaks due to the presence of both immature and mature individuals (Nakajima* et al. *2019, p. 145).*

Nakajima *et al.* (2019) further concluded, and Harland (2019) elaborated, that there were three stages in the aquaculture process: 1) fishing in places where natural wild populations of carp congregated during spawning (evidence = archaeological remains of multiple species, no evidence of water management); 2) marshy areas were altered through the digging of channels and management of hydraulics of ponds, and fish allowed to spawn and then were harvested (evidence = archaeological remains of single species, narrow range of body sizes and water management); and 3) substantial and constant husbandry, including food and specialised holding habitats, such as ponds and/or paddy fields (evidence = archaeological remains of single species, narrow range of body sizes, sophisticated water management and pond use). Harland (2019) added that Nakajima *et al.*'s Stage 3 was equivalent to domestication, because 'breeding, food and habitat are all controlled by people' (Harland 2019, p. 1379). Harland concluded that domestication of carp occurred independently in Asia and Europe. The research by Nakajima *et al.* (2019) suggested that Stage 3 domestication could be 5,000 years earlier than previously thought, and could be intimately linked to the development of wet rice cultivation. This blew previous thinking about the age of domestication of fish in general, and carp in particular, out of the water.

SPREAD OF CARP IN EUROPE

The Romans

The extant European populations of wild common carp evolved from the *C. c. carpio* subspecies, but as we have mentioned above, genetic evidence suggests that a bottleneck 10,000+ years ago considerably reduced the genetic diversity. This would seem to have been too early to be a result of the domestication of the European common carp, which if it were done by the Romans only dates back 2,000 years (Balon 1995a). Balon studied carp extensively for many decades, originally in situ in the River Danube in Bratislava (now Slovakia) in the 1950s and 1960s. He carried on with his research at the University of Toronto, and for many years at the University of Guelph and Axlerod Institute of Ichthyology (Pavlov *et al.* 2017). Balon's historical investigations, together with those of Canadian historian Richard Hoffmann, based at York University in Canada, have given us great insight into the spread of carp through Europe from pre-Roman, to Roman and then to medieval times.

Balon was convinced that the Romans not only exploited carp as a food source, but probably at least semi-domesticated it and facilitated its spread into Western Europe (Balon 1995a, b). First, he argued, Romans developed a taste for luxury during the last years BCE and the first years CE. This extended to food, and especially imported foreign food. Second, the Romans had breached the Alps and forged northwards to the River Danube, establishing military bases between modern Vienna and Budapest, with the Celts and Germans staring them down from the other side of the river. Although it is only 240 km from Vienna to Budapest, four garrisons were needed to maintain the peace; more than most other occupied regions. These garrisons included around 20,000 legionnaires, together with their retinue of wives, children, slaves and local tradespeople. So, the Roman-associated population of that region was many more than 100,000. And they all needed to be fed.

The fortresses and Roman towns occupied a region of the Danube that is known as the piedmont or foothills zone of that most majestic of European rivers that runs 2,850 km from the Black Forest in southern Germany to the Black Sea in Romania (Fig. 1.4). The piedmont zone is where the steeper Danube grades into the wider, more meandering Danube as it winds its way south and east. This part of the river also receives the waters of major tributaries: the Morava, Vah, Hron, Drava and Tisza rivers. Because of the river's location at this breakpoint in its slope, the area is a network of massive floodplains, some that have permanently and some seasonally flooded for millennia. To this day, restoration and conservation programs, like the International Commission for the Protection of the Danube River, are keen to realise the floodplain potential of the River Danube valley downstream of Vienna (https://www. icpdr.org/).

In Roman times, fortresses upstream and downstream of the Morava River were located at the upper end of one of the larger floodplains. Other fortresses were built at the downstream end of the floodplain. This floodplain was known as the westernmost extent of the spawning grounds of wild European carp. A hungry Roman population, an extensive floodplain full of big fat carp and an increasing appetite for the luxuries that fish can provide – all contribute to Balon's conviction that the time was ripe for wholesale exploitation of this abundant resource.

Excavations of some of these fortresses in the past few decades have found large numbers of fish bones, predominantly carp (Balon 1995a, 2004, 2006).

In the 1st century BCE, because of the Romans' increasing taste for luxury foods, saltwater reservoirs were built to keep fish alive and easily accessible. These *piscinae* soon became fashionable and ensured the supply of fresh fish regardless of the weather or the vagaries of fishing. *Piscinae* became a status symbol and wealthy Romans spent large amounts of money on such ponds, including freshwater ones although those were considered more plebeian than their saltwater counterparts. Ironically, this may have assisted in the westwards spread of carp, as troops – plebs by definition – retreated and took their taste for freshwater fish with them. Fish also began to be kept as pets and as ornamentals. Some fish were highly prized by their owners, in some cases bejewelled and mourned and buried when they died (Balon 1995a).

During the early part of the first millennium CE, there is evidence that the Romans bred freshwater fish in lakes and ponds (Zeuner 1963). So, it is possible that fish were not just held, but that what was done met a number of the criteria for domestication. We cannot be sure that Roman legionnaires, returning from the eastern provinces, brought carp back with them. But certainly, if they had chosen that particular fish, they chose one of the hardiest. The carp's tolerance of poor water quality – high temperatures, low dissolved oxygen levels and high salinity – means that it can survive and reproduce under conditions that are normally unsuitable for many other freshwater fish species. Furthermore, carp has a broad diet, and so can be fed a range of foods and survive periods of starvation.

After the collapse of the Roman Empire, Balon (1995b) hypothesised that, with the rise of Christianity and the establishment of monasteries, carp was kept and reared in monastic ponds. He speculated that wild carp were once more transported from the Danube westwards. The tradition of pond-rearing fish, he argued, was passed down the centuries – maybe not directly from the Romans, but who knows? The earliest written records of carp date from late 5th and early 6th centuries CE, when Magnus Aurelius Cassiodorus Senator (485–585 CE), secretary to King Theodoric (475–526 CE) and founder of the monastery Vivarium in Calabria, wrote of transporting wild carp from the Danube to Italy, 'from the Danube come carp and from the Rhine salmon. To provide a variety of flavours, it is necessary to have many fish from many countries. A king's reign should be such as to have foreign delegates believe that he possesses nearly everything' (Balon 1995a, p. 29). And around 1060, William of Hirsau from the Abbey at Cluny (eastern central modern France) produced a signing vocabulary that the monks could use during times of compulsory silence, which included *piscis qui vulgari nominee carpho dicitur* ('the fish which is commonly called carp') (Hoffmann 2023, p. 208). Hoffmann was sceptical that there were more than sporadic occurrences of carp in this part of Europe prior to the 13th century, because of the general lack of written and zooarchaeological evidence.

What's in a name? Well, quite a lot actually

The generic name for carp, *Cyprinus*, comes from the Greek *kyprinos* or *kyprianos*, a name used by Aristotle and derived from *Kypris*, a name for Aphrodite, the goddess of love (Balon 1995a).

Fig. 1.6. Frenzied carp spawning on the edge of Lake Hume, Australia, 2017. Source: Dirk HR Spennemann.

Balon speculated that it was so named because of the fish's fecundity and its propensity to indulge in noisy orgies of spawning in shallow vegetated waters (Fig. 1.6, Chapter 6). 'Carp' is a Celtic word that dates back to when the Celts lived along the shores of the River Danube in what is now Austria, Slovakia and Hungary. It was originally *charpho*, *carfo* or *charafo*, but eventually became *carpio*. The name was adopted by the Romans during their occupation of that region, and when they were displaced west it seems they took the fish and its name with them.

To add credence, if it is needed, to the presumed original wild distribution of the European carp, and its adoption and spread by humans during and after Roman times, all regions in which people of Celtic origin lived have *carp* as the root name for the fish (Balon 1995a) (Table 1.1). Regions where the carp was endemic all have their own local name for the species.

Water mills, fish culture and a changing medieval landscape

Prior to 1000 CE, humans had a relatively minor influence on rivers and floodplains in much of Europe west of the piedmont region of the Danube (Hoffmann 1995, 2005). It is true that humans had occupied these areas for more than a thousand years, hunting and fishing. But since the retreat of the Romans, populations had fallen and the subsistence agrarian communities scattered about the countryside had little impact on the natural environment. This changed dramatically over the next couple of centuries.

Table 1.1. Names for European common carp west and east of the westernmost extent of the presumed original distribution of wild fish populations (Balon 1995a).

West of River Danube piedmont zone		East of River Danube piedmont zone	
Language	**Name**	**Language**	**Name**
French	carpe	Hungarian	ponty
Italian	carpa	Serbian	sharan
English	carp	Bulgarian	saran
Spanish	carpa	Romanian	crap, ciortocrap, Ciuciulică,
Portuguese	carpa		ciuciu, ciuciulean,
German	Karpfen		olocari, ulucari,
Polish	karp		saran, ciortan,
Czech	kapr	Turkish	sazan baligi,
Slovakian	kapor		husgun
Danish	karpe	Croatian	šaran
Dutch	karper	Ukrainian	sharon, podrojek
Norwegian	karpe	Russia	sazan
Swedish	karp	Kyrgyzstan	kalynshyr
Finnish	karppi		

Sometime from about 1000 CE things started to change. Agriculture transformed largely because of technological advances and new methods of farming. The result was great improvements in production and ultimately, larger and more prosperous human populations (White 1962; Langdon 2002; Pounds 2014). Among the improvements were the use of horses instead of oxen as beasts of burden (owing to the invention of a superior horse collar), adopting a three-fold crop rotation system (planting twice per year, alternating with a fallow year) and, critically, the invention of a more robust plough that was able to cut deeper into the heavy clay soils of northern Europe, creating ridges and furrows for better drainage (White 1962; Andersen *et al.* 2016). These innovations increased food production and required more communal labour, both of which meant that communities grew substantially over time.

Hoffmann (1995) argued that the growing populations not only changed the landscape through increasing areas of land under plough, but through changes to hydrology as a result of deforestation, increasing nutrient and sediment inputs into rivers and putting greater pressure on fish populations because of the demand for food. Further, some scholars have referred to a medieval 'industrial revolution' (Carus-Wilson 2017) and, along with Hoffmann, argued that there was a dramatic rise in the number of water-powered mills, which meant that grinding grain, fulling cloth (cleaning, shrinking and felting), paper-making and many other industries could operate at speeds and produce quantities of goods at rates previously unthinkable. Hoffmann (1995) cited an example of the Aube River (a tributary of the Seine) in France, which in the 11th century had only 14 recorded dams but by the early 13th century had 200! The regions we now call France, and to a lesser extent Italy, seem to have been exceptional in the number and range of mill types built during this time, although water mills were widespread throughout many regions of Europe, China and the Middle East (Lucas 2005; Brykała and Podgórski 2020).

Water mills alter riverscapes in profound ways that affect fish, sediment and plants (Wood and Barker 2000; Lenders *et al.* 2016; Brown *et al.* 2018; Brykała and Podgórski 2020). For a mill to function, the wheel must be turned by moving water, usually from a spring or small or large river, but sometimes associated with a lake outlet. For the water to have enough wheel-driving force, a head of water must be built up above the wheel in a reservoir (the millpond) and the water must be channelled through a narrow passage (the mill race) which causes it to fall across the wheel. In running water, a dam had to be built to create the millpond and mill race in which the mill wheel was placed. Creating a pond from an existing flowing water system results in a water body with larger surface area and much slower flow. This has several effects: the slow or still water means that suspended particles drop out of the water column and sediment builds up in the mill dam; the water is clearer than it would have been otherwise because lower suspended sediment means that more light penetrates the water and so enhances aquatic plant growth; and the greater surface area exposed to sunlight means that the water is warmer, especially in the upper part, which further enhances the growth of plants and other organisms (Hoffmann 1996).

At this time, and indeed for millennia, warm-water fishes dominated the faunas of eastern Europe and the Balkans (Maitland and Linsell 2006). Many of these were cyprinids, including, as we have seen, the common carp. These species tend to prefer sluggish lowland rivers and can complete their life cycles wholly in freshwater. By contrast, salmons and trouts (Salmonidae), whitefish (Coregonidae), European sturgeon (*Acispenser sturio*), shads (*Alosa alosa* and *Alosa fallax*), barbel (*Barbus barbus*), bream (*Abramis brama*), roach (*Rutilus rutilus*), minnow (*Phoxinus phoxinus*) and gudgeon (*Gobio gobio*) dominated western European faunas. These species tend to prefer cooler, faster-flowing streams, and importantly, along with lampreys (Petromyzontidae) and European eel (*Anguilla anguilla*), the salmonids and sturgeons migrate between freshwater and the sea as a part of their life cycle (Jenkins 1946). The replacement of continuous stretches of flowing water with alternating millponds and dams with mill races meant that fish movements were restricted. This was a serious problem for migratory species. Furthermore, the mill dams concentrated flow, which provided ideal opportunities to catch fish moving downstream or attempting to move upstream (Hoffmann 1995).

Hoffmann (1995, 1996) disagreed with Balon's (1995a) hypothesis about the spread of carp westwards during Roman times (although this could have happened at least twice in history in that part of the world), and instead suggested there were three phases to the expansion of the carp's range from its natural distribution:

> *The first phase between perhaps the seventh century and the eleventh principally carried the fish up the Danube and into at least some west-flowing tributaries of the middle Rhine. Some [archaeological] remains also indicate carp in waters north of the Danube. In a second phase (twelfth through early fourteenth century) carp radiated across most of the economic and cultural heartland of medieval northwestern Europe, from the lower Rhine/Mass region south to the Paris basin and Burgundy. A final stage, probably occurring after the mid-fourteenth century shift of economic trends, extended common carp into the outer periphery of the west, an arc from southern Scandinavia into east central Europe. Though carp*

> *are recorded in Italy at the end of the Middle Ages, the diffusion pattern at no time accords with radiation from there (Hoffmann 1995, p. 72).*

Hoffmann (1995) further concluded that there is no evidence of fish culture being part of the process of carp range expansion until the final stage.

Although some of the details are sketchy and there is some disagreement among historians, it is highly likely that the construction of mill dams, the rise in pond fish culture, major and widespread changes to the countryside during this period in history and the undoubted human-assisted movement of carp resulted in establishment of this species in western and northern Europe, including Britain towards the end of the Middle Ages (Hoffmann 2023; Chapter 2), well beyond its natural range. It joined a range of still-water-adapted fishes, such as pike, and the cyprinids roach, bream, tench and dace, that were already abundant. Through the feeding actions of carp – which disturb plants and sediments and increase turbidity – the species probably modified its new habitat to the detriment of other native fishes. By the 14th century, the fish-eating public highly ranked the culinary qualities of pond-reared carp.

The rapid rise in western and northern European human populations and the significance of fish as a meat substitute because of religious dietary restrictions meant that freshwater fish stocks were overfished (Hoffmann 1996). Migratory, cool-water species suffered the most, not least because of their increasingly limited movements and alterations to spawning areas because of mill dams, but also because for hundreds of years their presence on dinner tables was a status symbol. Their numbers dwindled over time. Zooarchaeological studies indicate that sturgeon from 17 southern Baltic sites made up 70% of fish consumed in the 7th–9th centuries, but only 10% in the 12th and 13th centuries (Benecke 1986). The situation was equally dire for salmons, trouts and whitefish, which by 1500 CE had become a rare commodity even for those who could afford them.

Carp, and to a lesser extent, eels (which can survive for decades being landlocked), on the other hand, thrived under the altered conditions (Fig. 1.7) (Hoffmann 1996). By 1400, carp was the most prominent fish served to diners in Paris, Orleans and La Charité-sur-Loire (Hoffmann 1996). Archaeological and written evidence indicates that in many rivers in western Europe, warm-water fish species, particularly carp, adapted to still or slow-flowing conditions. They took over where cool-water, migrating species had previously dominated. Although small to medium rivers were affected the most by mill development, even large rivers – which suffered death by a thousand mill-dams in their tributaries – were impacted. Hoffmann (1996) summarised the situation thus:

> *Characteristic differences between the fishes that rose in importance and those that fell thus argue for human impact on medieval aquatic ecosystems more complex than can be ascribed to overfishing alone. The varieties that increased in dietary importance and apparent natural success may be characterised by greater tolerance, even preference, for warmer, more turbid, and quieter waters (Hoffmann 1996, p. 652).*

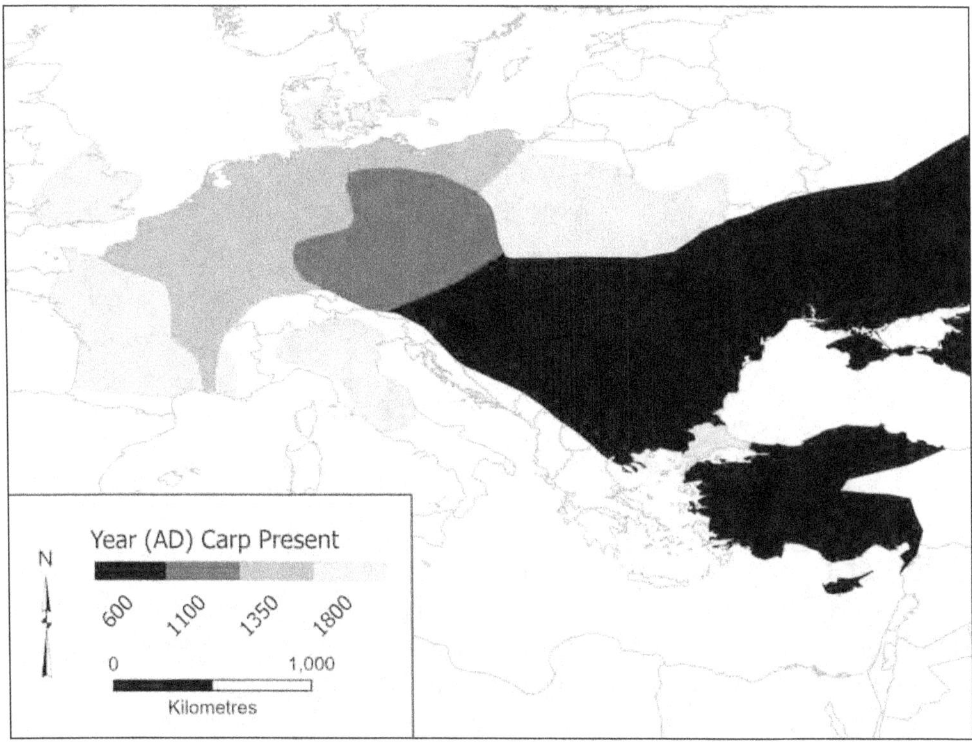

Fig. 1.7. The westward spread of the common carp in medieval Europe. Source: Redrawn after Hoffmann (1996).

*****

So, by around 1600, the common carp had reached most of western Europe. Carp and similar species had replaced the once abundant trouts, salmons and sturgeon, and had come to dominate the transformed landscape of the region. As we will see in the next chapter, the 'Carpocene' – the era of the carp – was just around the corner. It resulted from the rise of aquaculture and the popularity of fish-eating in western Europe, the encroachment of carp into that part of the world, the rise in Christianity and a particular monkish fondness for ponds. It was inevitable that carp, its aquaculture potential well and truly tested in continental Europe, would end up in England and be a hit there as well. The rise of carp in England would get a further boost from none other than Henry VIII and the Protestant Reformation.

2

Fish culture in Europe and the Carpocene

Your bait of falsehood takes this carp of truth (William Shakespeare, Hamlet, *Act 2, Scene 1).*

WHETHER CARP ACCOMPANIED THE RETREATING ROMANS 2,000 YEARS AGO or its range expansion took place somewhat later is open to debate, as we saw in Chapter 1. But there is compelling evidence that a medieval agrarian revolution and the changing landscape of Europe played an important part in modifying creeks, streams and rivers into ideal environments for carp to live, breed and grow. It was probable too that the rise of Christianity and its associated religious dietary restrictions, the construction of monasteries and monks' special fondness of fishponds increased the popularity of carp in western Europe. Although we know something about how, when and at whose hands carp were introduced into Australia in the second half of the 19th century, and that influential people had been advocating on the fish's behalf for some time, the story behind why this species – unlike the trouts and salmons – was chosen for introduction has received much less attention. Part of the answer undoubtedly lies in the biology of the species. But part of the answer also lies in the development of aquaculture from the early Middle Ages, first in small artisanal enterprises which held and reared fish for local consumption, and later as a sophisticated and much larger-scale industry in western Europe. Carp played an important role in the success of aquaculture once it had reached that part of the world. With its introduction to England not long after, carp's popularity soared to dizzying heights. It even made an appearance in two of Shakespeare's plays! A fish by any other name ...

In this chapter, we describe the development of the culture of fish in Europe, its origins and changing nature as technologies improved over the centuries. We highlight the carp's leading role in this industry, and follow the fish as it is introduced to England. Monasteries and monastic ponds provided the foundations on which the English aquaculture industry was built, but it was catalysed by Henry VIII's Dissolution of those monasteries during the Protestant Reformation. The fishponds fell into the hands of wealthy landowners, and from humble beginnings a thriving English fish culture industry ultimately grew. Our aim is to set the scene of early 1800s England, with carp waiting in the wings of its watery stage. In doing so, it will become clear why this species was a prime candidate for introduction by Australian acclimatisation societies. It was they who unleashed carp into the unsuspecting Australian landscape.

FISH CULTURE IN EUROPE

In Europe, fish became an important source of food for Christians following the decline and fall of the Roman Empire. For many days during the year (up to 250 in some regions and denominations, if Fridays and the 40 days of Lent are included) only crustaceans, shellfish and fish were permitted to be eaten (and, it seems, unborn rabbits: Balon 1995b). Although fish were generally obtainable from rivers and lakes, a reliable and abundant source was indispensable. If, as Eugene Balon speculated, Romans had conducted a form of fish-keeping, however rudimentary, monks may have followed suit. And if the need was for a hardy, fast-growing, tasty fish, carp was the obvious choice in many of the warmer regions of Europe. Monasteries sprang up in western Europe from the late 4th century, arriving via the Middle East. Early on, monasteries were relatively small and reliant on the generosity and patronage of local people but they later acquired large tracts of land, sometimes including ponds, and monks began holding, and later breeding, freshwater fish. Technically, domestication of carp – and other fishes – was a while off yet, but its roots were here.

It was not until the 13th century in Europe that written accounts of breeding carp in captivity appeared (Hoffmann 1996, 2005, 2023). Balon (1995b) suggested that the lack of written documents does not preclude the possibility that this happened earlier – it might just have been too common a practice to warrant description. From the 1500s, more and more writing appeared on the subject of culturing carp in ponds (Hoffmann 2023). And although some writers mentioned Roman practices in bygone days, those allusions may have been more fanciful than factual (Hoffmann to Balon 1995, pers. comm.). But by no means was it just carp being cultured. Indeed, we need to take a step back and look at the context behind the rise and purpose of fishponds and fish culture in Europe more generally. This ultimately led to the commercialisation of carp in Europe and England as a table fish, and contributed to the motivation that resulted in its eventual introduction to Australia.

Origins of European fishponds

Along with dams constructed for milling purposes, the early medieval period saw a proliferation of fishponds, especially associated with monasteries, with both the practicalities of food production and aesthetics in mind (Fig. 2.1) (Hoffmann 1995, 2023). Many monasteries were too far from the sea to guarantee unspoiled transport of saltwater fish, and while expensive imported fish were popular if they could be obtained, preserved fish were not, especially among the religious and secular upper classes. The species of fish being bred, cultured or simply grown out very much depended on the region, and especially the climate (Bonow *et al.* 2016b). For example, in the non-Scandinavian countries of Poland, Germany, France and Italy, since medieval times common carp was generally the most popular aquaculture species. In regions of higher latitude, however, the harsh winters meant that the more cold-tolerant crucian carp dominated (Bonow *et al.* 2016a; Bonow and Svanberg 2016; Cios 2016; Hofmeister 2016; Hufthammer and Moe

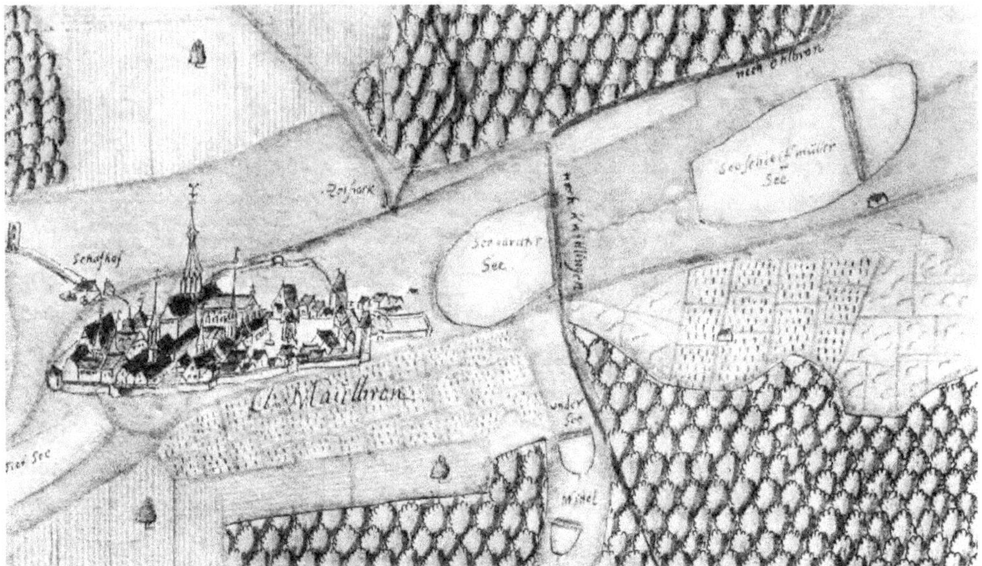

Fig. 2.1. Monastery fishponds [the two 'Sees' to the right of the town], Kloster Maulbronn, Germany. Source: Kiesersche Forstkarte Nr. 95 from Stromberger Forst von 1684, Altwürttembergisches Forstkartenwerk des Andreas Kieser, CC.

2016). Other species, such as pike (*Esox lucius*), perch (*Perca fluviatilis*), bream (*Abramis brama*), roach and tench, were also popular at times. Indeed, fishponds and early aquaculture in much of Europe pre-dated the spread of common carp to those regions.

As we have seen in Chapter 1, fishing and eating fish goes back many thousands of years in human history. As Hoffmann (2023) remarked, where fish consumption is concerned, fishing in rivers or lakes for wild stock is at one end of a continuum and fish farming is at the other. But a lot has to happen in between. And for fish farming to have longevity, there has to be a happy marriage between technology and suitable fish species. Understanding domestication *per se* is not our goal here. Instead, we aim to understand the processes associated with the development of aquaculture in Europe, in particular, because it was from England that carp were introduced to Australia in the latter part of the 19th century.

Genuine fish culture entails human manipulation of all the same variables as agriculture – mobilising land, labour and capital to produce biomass for cultural purposes, normally human food. Managers select specific varieties for production and specific individuals for reproduction (breeding stock). Reproductive isolation drives evolution of domesticated varieties, genetically distinct from wild populations. Habitat is closely controlled to maintain supply and quality of water and of energy (nutrients). Environmental variables such as predators, pests and pathogens are monitored and combated. The aquaculturalist's work to manage inputs and outputs more resembles that of a shepherd or grain farmer than that of a fisher trying to find and gain possession of an animal in the wild (Hoffmann 2023).

For hundreds of years before true aquaculture, people kept fish alive in tanks, millponds and moats so they would remain fresh by the time they reached the table or market. By the mid-1500s, the process was much more sophisticated and knowledge of the biology of the species being cultured was more comprehensive. European aquaculture likely had its origins in what is now France in the 11th and 12th centuries, with monks and the nobility. Ponds may have originally been constructed to control drainage, but an enterprising abbot may have seen the potential for throwing local fish into the pond and catching them at a later date to provide a solid tasty meal, especially at times when meat was forbidden. Typically, ponds contained only one species of fish. The keeping of multiple species came later as fish husbandry grew in sophistication.

When pond construction and ownership were in their infancy, there was often sharing of the resource: when the pond had water and fish in it, it belonged to the abbey; when the pond was dry, it belonged to the nobleman and a crop was sown and harvested in that space. The fish produced were often shared between those two parties. The techniques of pond construction and fish keeping by the Normans in continental Europe made their way over the Channel with the Norman invasion in 1066, and were adopted in England. By the way, William the Conqueror himself established a fishpond in 1086–1089 at York, when a dam he had built to protect his castle from the River Foss created a *stagnum*, or large pool, in which he stocked bream and pike (Bond 2016b).

Development of aquaculture and advent of the Carpocene

The most common fish in fishponds in the middle medieval period were bream and pike. In the mid-12th century, carp had only made its way to the main stem of the middle Rhine. But the carp explosion – what we call the Carpocene – was not far away. Within a century, by 1260, carp had reached northern France and the Low Countries (the Netherlands, Belgium and Luxembourg), was stocked in French ponds and sold in Parisian markets. Fishpond technologies and fish husbandry had developed to such a level of sophistication that the time was ripe to take advantage of this fishy invader. There was also enough knowledge of carp biology for its production in the pre-existing fishponds to be successful.

By the turn of the 13th century, carp keeping and breeding were widespread in northern and central France. Most estates of any size, whether monastic, royal or noble, had fishponds, and those fishponds contained and grew carp. The main purpose of the fish was to supply the tables of the people who grew them, but surpluses went to markets or to stock neighbours' ponds. The system of fishpond management and fish breeding and rearing, developed in northern France, soon radiated south, north and especially east – particularly to what are now known as central Germany, Czechia, Austria and Poland.

There are particularly good aquaculture records for Poland, which until recently were all in Polish and had a relatively limited audience (Cios 2016). Before 1400, however, there was relatively little written about aquaculture in that country. The royal accounts of Krakow from 1388–1420, for example, mention eel (*Anguilla anguilla*), river lamprey (*Lampetra fluviatilis*), catfish (*Silurisglanis*), grayling (*Thymallus thymallus*), Atlantic salmon (*Salmo salar*), roach, bream, sturgeon (*Acipenser oxyrinchus*), burbot (*Lota lota*), ide (*Leuciscus idus*), stone loach (*Barbatula barbatula*) and pike but there is no mention of carp. Yet 100 years later, carp was well established

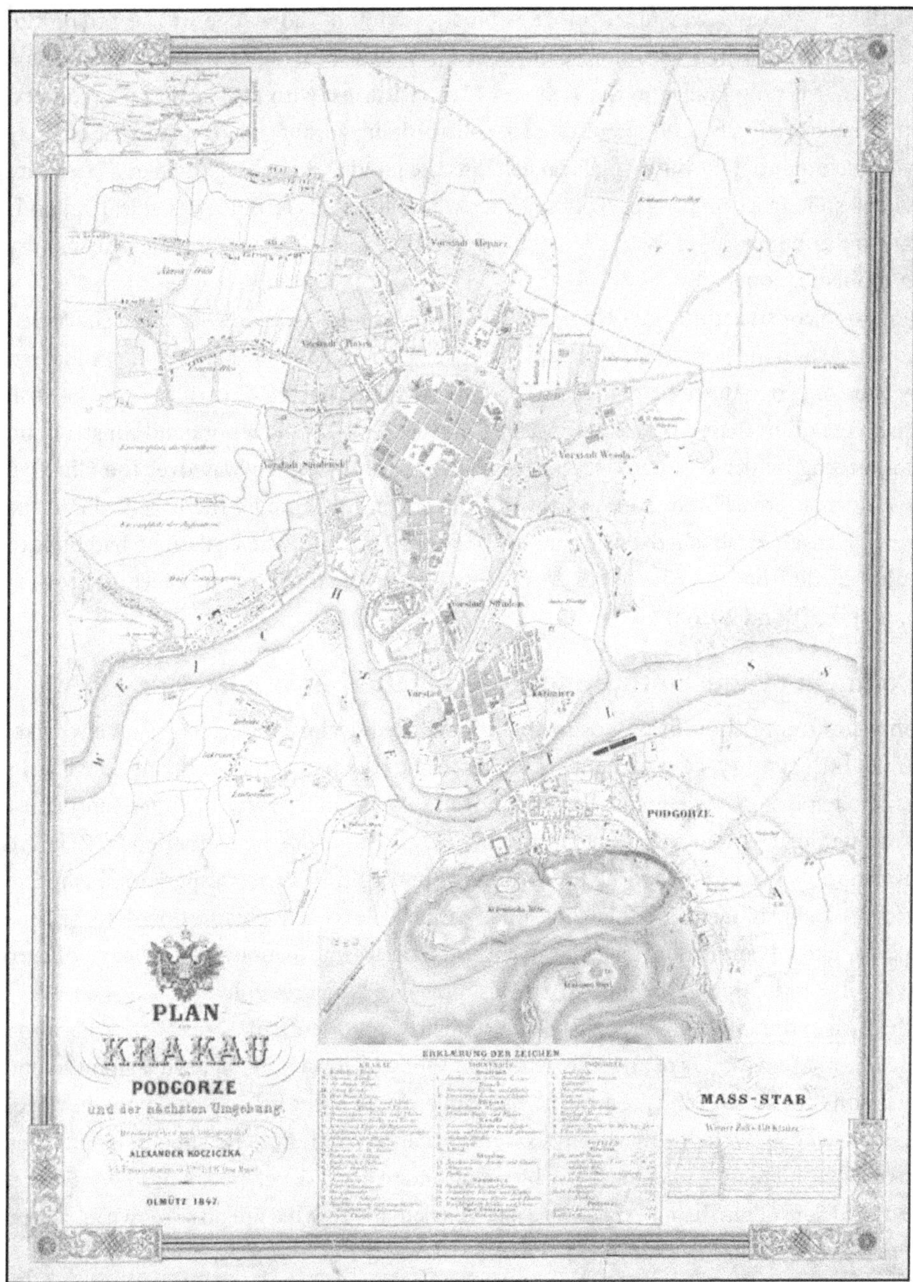

Fig. 2.2. Old map of Krakow, Poland. Source: Alexander Kocziczka, Public domain, via Wikimedia Commons.

in Poland, probably brought in through Germany or Moravia. The local climate was conducive for carp rearing (though not ideal for spawning), there was proximity to the huge market at Krakow and carp could be shipped down the Vistula River in wooden boxes, all of which added up to the region of Malopolska becoming the centre of Polish carp production from the 15th century (Fig. 2.2) (Cios 2016). Indeed, a mild climate, together with a nearby river to supply water for ponds and transport of fish, seem to be the general requirements for aquacultural success during this period throughout Europe.

Carping around Scandinavia

In Denmark, freshwater fish production began in the Middle Ages, and common carp and crucian carp dominated (Hofmeister 2016). Nowadays, rainbow trout (*Oncorhynchus mykiss*) has effectively replaced both species in popularity. The first mention of a 'fishpond' in Denmark was in 1241, although little was known about the nature or purpose of such a pond. Danish royal accounts from the mid-15th–16th centuries failed to mention common carp, despite many other species being included. But in a relatively short time, carp was being extensively farmed throughout the country. As the popularity of saltwater fishes declined in the 17th century, carp farming intensified and the Danish royals and Danish nobility got into the act of commercial fish farming. The King of Denmark even brought in German experts to advise him on the latest fish-farming practices.

In Sweden and Finland, aquaculture has mostly involved crucian carp, because the cold winters are not conducive to the survival of common carp (Bonow and Svanberg 2016). Norway, on the other hand, has seen more farming of common carp over the centuries (Hufthammer and Moe 2016), as have Latvia and Lithuania (Bonow *et al.* 2016a). Rainbow trout is a more popular aquaculture species than common or crucian carp in Estonia.

The fortunes of the aquaculture industry in northern Europe waxed and waned and sometimes waxed again, because of technological advancements (e.g. transferring fry from breeding to growing-out ponds), broader economic patterns (e.g. high wheat prices making agriculture more profitable at times than aquaculture), changes in the public's taste for particular types of fish (a preference for saltwater and freshwater fish seems to alternate historically, driven by palate or by status or both) and, perhaps more surprisingly, because of wars (Bond 2016b; Bonow *et al.* 2016b). Wars in at least Poland, Denmark and England, and no doubt other countries, especially during the 17th century, mobilised aquaculture workers as well as many other men. Probably more significantly, wars brought hordes of hungry soldiers marauding through the countryside needing lots of food to fill their stomachs so that they could fight, and so many fish farms were drained and their fish stocks looted (Cios 2016; Hofmeister 2016). Added to that was a general relaxation of religious dietary restrictions, partly due to the Reformation (see below) but mostly because men were not particularly effective fighters if they had to fast for extended periods during military campaigns.

CARP IN ENGLAND
Fishponds in England

Although there is no universally agreed date regarding the introduction of carp to England, it likely took place just before or around 1350, perhaps on the eastern seaboard of England in East Anglia, and possibly via the Netherlands (Currie 1991). At that time, and as far back as when the Norman aristocracy ruled Britain, fishponds were the privilege of monasteries, nunneries and priories or wealthy landowners, knights and royalty. They symbolised status through the ownership of land (*demesnes*, land attached to a manor house) and the management of water and the bounty they provided (Currie 1990; Dyer 2003; Bond 2016a).

There could be many motives for constructing fishponds: to fulfil fundamental subsistence needs; to contribute variety to a cereal-, vegetable- or meat-based diet; to reflect the status of an owner able to indulge in conspicuous consumption; to provide revenue from commercial sales; to promote aesthetic enjoyment as a feature of garden design; and to facilitate the enjoyment of angling as a pastime (Bond 2016b).

Monastic fishponds flourished from the middle of the 12th century, when benefactors wanting to curry favour often donated existing ponds for monks' use. Alternatively or as well, monasteries were allowed special fishing rights to ponds owned by benefactors (Currie 1989). The ponds typically kept perch, roach, bream, tench and pike.

For the construction of fishponds, streams had to be altered as they were in the construction of mill dams (Bond 2016a). In some cases, whole rivers were diverted and channelised, dams built and land either side flooded. This was not altogether popular with those affected, but abbots mostly got their way. The size, status or wealth of a monastery did not necessarily correlate with the number or extent of fishponds. Some large monasteries had few ponds, whereas others, like the small Augustinian priory at Maxstroke, had 10 ponds and a moated enclosure (Bond 2016a). Some ponds reached such large sizes (e.g. 53 ha) that a boat and nets were needed to extract fish. The lowest of a series of ponds may have had a mill race and mill attached. Fish were stocked in ponds sometimes through royal grants, sometimes through transfers between monasteries, but often by direct purchase. Pike, perch, roach, tench, bream and eels were bought from professional fishermen and transported sometimes considerable distances. The types of fish kept needed to be tough to withstand a potential eight- to 10-hour trip.

Monastic restrictions on diet, as we have described, meant that fish were an important source of food for many days in the year (Bond 2016a). Contrary to Hoffmann's (2023) way of thinking, Bond (2016a, b) argued that marine fish were readily accessible and cheap, even far inland in England, and were the preferred choice of monks. He also argued that monastic fishponds existed not to provide fish for everyday eating, but mostly to produce fish for meals for important visitors and feasts. But the size of the ponds and the number of fish stocked, suggest more fish were kept than would be needed only for special occasions. Whatever the reason, fishponds remained a common feature of monastic communities until the Dissolution that began in 1536 under Henry VIII. The Dissolution of more than 600 monastic communities was linked to the Protestant Reformation sweeping through Europe during the 16th century. Students of history will know that it was part and parcel of Henry's political and religious intercessions in England, his desire to annul his marriage to Catherine of Aragon so he could marry Anne Boleyn, and the general dissatisfaction with monastic excesses and power. There was not, however, overwhelming public support for the Dissolution. Nevertheless, it went ahead, and dissenters – common or noble, secular or religious – were suppressed or executed. Much of the monasteries' wealth was redirected to the royal purse and the land and buildings were typically bought by nobles and others of high status, aiming to further increase their

Fig. 2.3. Hindringham Hall: The stew [fish] pond. Source: Lynda Tucker.

status. In 1538, for example, 300 carp were removed from the ponds of the dissolved London Charterhouse, a Carthusian priory founded in 1371, and 100 fish were sent to none other than Henry VIII himself (Bond 2016a).

In England, the fishpond became a secular status symbol, harking back to its pre-Christian origins (Fig. 2.3). Freshwater fish were particularly prized because they had to be 'kept' (Maccarinelli 2020), whereas sea fish were obtainable by all (Galloway 2017). Freshwater fish not only served a practical purpose as a highly prized food, but they and the ponds in which they swam also served an aesthetic one, for the owners themselves and to show off to visitors. The significance of fishponds as ornamentation and for angling for pleasure dates back to the 12th and 13th centuries (Bond 2016a; Maccarinelli 2020). The ponds in some monasteries may also have been used to grow aquatic plants for medicinal purposes or as features to give meditative pleasure to monks recovering from illnesses (Harvey 1992). Manor houses, medieval lodges and monasteries were sometimes built with viewing terraces, large windows or aspects that gave spectacular water views for anyone passing by or visiting (Taylor 2000).

The aristocratic owners of fishponds were typically not focused on the commercial production or sale of fish, but on supplying their own demands (Currie 1991). They also rarely leased the ponds, because the rent received was negligible. The ponds were treated as 'underwater pasture': left to themselves until they were drained to retrieve the fish that had grown fat eating whatever was in the ponds (sometimes supplemented with food thrown in) and that now flapped around at the bottom. River fish were in higher demand than pond-reared fish, probably because rivers were more productive, yielded different species of fish and were less easily controlled and policed. Pond fish were, however, reliably present and something to show off. Pond angling later became a popular diversion, particularly for young people bored with the normal games of cards and billiards (Currie 1990; Bond 2016b).

By the 14th and 15th centuries, demesne farming was less popular, and ponds were increasingly leased to people of lower economic means and social status. Freshwater fish were still considered upper-class fare, and their possession and consumption was a sign of privilege. In the 1360s, on the bank of the Thames at Southwark in London there was an arrangement of plots, called 'The Stews' (Currie 1991). These were a series of ditches managed by fishmongers, in which fish were kept and fed, then sold at market. But stews were by no means unique to London. By the 15th century there was an extensive range of fishponds throughout England, formerly managed by wealthy landowners but increasingly run as commercial enterprises. The fish reared in those ponds were sold to a burgeoning market, most notably fashionable Londoners willing to pay a decent price.

Introducing carp to England

Currie (1991) speculated that it was probably no coincidence that carp came into England around the late 14th century, during the shift in the ownership and management of fishponds and the expanding freshwater fish economy. Hoffmann (2023), however, was somewhat sceptical about the date and suggested a later introduction. Regardless, by the mid-1500s, carp had spread throughout the country and become the most popular freshwater fish in England.

> *The reasons for its [the carp's] introduction as a popular species must now start*
> *to become plain. The ... secularized fishing industry of England, in an age of*
> *increasing materialism, could not have failed to notice the opportunities awaiting*

> *it ... They would have been quick to notice the hardiness of the carp, the ease
> with which it could be transported, and, more importantly, its rapid growth rate
> compared with other freshwater species. As fish culturalists know only too well, the
> quicker a fish will grow, the more money there is to be made (Currie 1991 p. 101).*

Currie also thought there may have been a deliberate commercial intent in bringing carp into England at that time (Currie 1991). Although it is still uncertain exactly when and from where they were introduced, as we have stated, royal kitchens mention carp for the first time in the 14th century (McDonnell 1981) and Currie suggested they may have been introduced from Flanders, then the home of some of the most experienced fish farmers in Europe (Currie 1991; Hoffmann 2023). Being adjacent to the Low Countries, East Anglia had a long tradition of connection and trade with that part of Europe, and English writers advocating the keeping of carp mentioned the work of Dutch fish husbandry experts. A Dutchman, Fleming, was a lessee at The Stews in London in 1381. Maybe a coincidence. Maybe not.

The first written record of carp in ponds in England dates from East Anglia in 1462, when the Duke of Norfolk stocked his ponds with the fish (Currie 1991). Between 1462 and 1472 he placed 800 carp in multiple ponds, and in 1465 the Duke gave carp to five of his neighbours so that they could share his good fortune. By the 1530s, it seems that carp were abundant and widespread in England, north, south, east and west, as testified by many writings on the topic. The Prior of Llanthony Secunda, Gloucester, in the Cotswolds, sent some carp to Henry VIII in 1530. In 1537, Thomas Wriothesley put 500 carp in four previously monastic ponds, each a mile in length, at Titchfield in Hampshire. And in 1538, a letter to Oliver Cromwell mentioned carp being given as a present to a Lady. By the turn of the century, the nature of carp, its hardiness and wiliness when being angled, had become so established in English culture that Shakespeare mentioned them twice (Spates 2013). In *Hamlet* (circa 1600):

> *There was he gaming, there o'ertook in's rouse,*
> *There falling out at tennis,' or perchance*
> *'I saw him enter such a house of sale' –*
> *Videlicet, a brothel – or so forth. See you now,*
> *Your bait of falsehood takes this* **carp** *of truth*[1] *(Hamlet, Act 2, Scene 1).*

In *All's Well That Ends Well* (circa 1603–1606):

> *[FOOL] Foh, prithee, stand away. A paper from Fortune's close-stool*
> *to give to a nobleman!*
> *[Enter LAFEW]*
> *Look, here he comes himself. Here is a purr of*
> *Fortune's, sir, or of Fortune's cat – but not a*
> *musk-cat, – that has fallen into the unclean fishpond*
> *of her displeasure and, as he says, is muddied*
> *withal. Pray you, sir, use the* **carp** *as you may,*

---

1 Spates explains it thus: 'Polonius, caught up in his cleverness, introduces his carp conceit as an example of a wary creature that needs to be angled cleverly with "the bait of falsehood"' (Spates 2013, p. 1).

for he looks like a poor, decayed, ingenious, foolish,
rascally knave. I do pity his distress in my
*similes of comfort and leave him to your lordship[2] (*All's Well That Ends Well,
Act 5, Scene 2).

Our point in including these two quotes from Shakespeare's plays (Spates' motivation is the same, pre-dates ours considerably and with more erudition) is to emphasise the fact that by the time of writing *Hamlet* and *All's Well That Ends Well*, 50 years before the publication of Izaak Walton's *The Compleat Angler* in 1653 – the universally acknowledged bible of early angling – the habitats and nature of carp were as familiar to all English people (or at least Shakespeare's intended audience) as they were on continental Europe centuries earlier.

By the time that Shakespeare was writing his masterpieces, carp had established itself as the most popular freshwater fish in England, according to John Taverner in his *Certaine Experiments with Fishe and Fruite*, published in 1600, and Robert North in *A Discourse of Fish and Fishponds*, published in 1713 (both cited in Currie 1991). Carp was now prized not just for eating but also for angling, and enough individuals had escaped their pond-ish confines for carp to be considered both a river and pond species. A plethora of fish-keeping and fish-catching handbooks and pamphlets followed (e.g. Walton 1653; Balgrave 1669; Worlidge 1669; Smith 1673; Mortimer 1707; Bradley 1721; Hale 1756; see Currie (1991) for bibliographic details), with carp praised above all other fishes.

Carp had well and truly replaced bream, pike and tench, formerly the most popular freshwater fishes eaten by royalty and the aristocracy. And it is not hard to see why: it was extremely hardy and, as Currie's quote above mentions, grew faster than its finny compatriots. Whereas it might have taken bream five, six or even 15 years to reach table size, carp was ready in as few as three. And it needed relatively little space, would eat most things given it, was tolerant of poor water quality and could be transported relatively easily to market. The reasons for carp's popularity previously with the Chinese, Romans and medieval continental Europeans meant that its popularity in its new-found English home was assured. This peculiarly English popularity lasted almost 400 years.

*****

With the development of sophisticated aquaculture practices and the happy coincidence of carp dominating the transformed riverscapes of western Europe, the Carpocene had been underway in Europe for centuries, with England lagging a little behind. Carp had established itself in England and been well-known and well-loved for almost 200 years by the time that Australia was colonised by Europeans. As we will see in Chapter 3, in the mid-19th century, gold was

---

2 We like this one even more, although it takes a bit to appreciate. 'In this instance, Parolles, dirty and repudiated, comes begging to be reintroduced into the court. Lavatch opens with a play on words based on free association between excrement and fishponds overlaid with images of flesh and fish: Parolles' written request is toilet paper, and Parolles himself is Fortune's toilet. Playing on the idea of *pur* as a term for excrement, Lavatch essentially calls Parolles the shit of Fortune but then creates a lexical shift, so *pur* is reconceived as the 'purr' of a cat. Lavatch clarifies this is no musk-bearing civet but a poor cat that has fallen into an "unclean fish-pond", which allows for yet another semantic shift from flesh to fish – wherein Parolles becomes a carp, the only fish that could dwell in such a polluted environment' (Spates 2013, p. 3).

discovered not far from a young Melbourne, Victoria, and within a decade the colony was transformed beyond recognition. The population skyrocketed as people arrived from all around the world to try their luck on the gold fields. Although gold mining was an environmental catastrophe, the wealth and population catalysed by the discovery of gold meant that the colony grew in leaps and bounds, and so did the arts and sciences. Imported scientists and scholars in many cases recognised the enormous damage wrought by gold mining. They jumped at the chance to make astounding 'discoveries' and establish successful careers in this far-flung part of the world, but some also sought to rectify the environmental damage through the importation of hardy and 'useful' species. Scientific societies, many styled on early European ones, blossomed like flowers on a river red gum. One type of society – acclimatisation – is of particular importance to our story on the introduction of the carp to Australia. And it is to acclimatisation that we turn to in the next chapter.

3

Acclimatisation: Its origins and motivations

The rivers, lakes, and seacoasts once restocked, and protected by law from exhaustion by taking fish at improper seasons, by destructive methods, and in extravagant quantities, would continue indefinitely to furnish a very large supply of most healthful food, which, unlike all domestic and agricultural products, would spontaneously renew itself and cost nothing but the taking (Marsh 1864, p. 118).

REPORTS, SCIENTIFIC PAPERS AND THE OCCASIONAL BOOK CHAPTER USUALLY include a summary sentence on the fact that carp was introduced in a particular year and perhaps mention that the Geelong and Western District Acclimatising Society or the Acclimatisation Society of Victoria was responsible (Shearer and Mulley 1978; Koehn 2004; Brown *et al.* 2005). The accounts may go on to say that the first introduction or two did not work, because the fish did not survive. But bear with us, as our story takes us once more back to France and England and to the rise of acclimatisation societies, who founded them and what their motivations were. This is because *acclimatisation* and *acclimatisation societies* are an integral part of the story of how carp got to Australia. This chapter will explain why in 1862 Professor Frederick McCoy, director of the National Museum of Victoria, stood up at the inaugural meeting of the Acclimatisation Society of Victoria to mention the introduction of carp and other 'pond' species to Victoria and, a few years later, extolled the virtues of continued introduction of the common carp – and, most decidedly, not other inferior species of carp – to Victorian waters.

Acclimatisation societies generally get a bad rap from Australians with any knowledge of history, although we suspect that many Australians may never have heard of these now largely defunct organisations. It is a chapter of Australia's history that has been considered shameful – not surprisingly – and we are still dealing with its legacy more than 160 years later. After all, acclimatisation societies introduced plants and animals that have done untold damage to the Australian environment over many decades (Rolls 1969; Low 2001). They introduced foxes, brown and rainbow trout, deer, goats, blackbirds, thrushes, sparrows, common mynas, starlings, scotch thistle, privet, hawthorn, gorse ... The list goes on and on and on. Their misguided introduction of many species ranks them pretty high on the list of Australia's historical environmental vandals. Although there were acclimatisation societies (and indeed other scientific and cultural societies) in New South Wales (1861: Strahan 1992), Queensland (1862: Pearn 2020), South Australia (1878) and Western Australia (1896), it was in Victoria that acclimatisation – and the introduction of carp – were most enthusiastically promoted and pursued.

A plunge into the history of Victoria is needed to make sense of the Acclimatisation Society of Victoria, because the story is more complicated than it first appears (Minard 2019) – especially how scientists, government and the general public perceived, responded to and then approached solutions to the environmental devastation wrought by the gold rush of the 1850s (Frost 2013). We need to understand what the initial French *acclimatation*, that became *acclimatisation*, meant when first proposed, in particular in France and England and later in Australia and New Zealand (Gillbank 1986). We also need an understanding of the taxonomy and classification of animals that existed at that time, because these ideas influenced the ethical attitudes of scientists introducing 'new' species to foreign shores (Anderson 1992).

Furthermore, the mix included, as always, politics, egos and scientific territories. Scientists, like foundation Professor of Natural Science at the University of Melbourne, Frederick McCoy, and Government Botanist and Director of Botanic Gardens, Ferdinand von Mueller, rejected Darwin's revolutionary evolutionary ideas and sought to prove him wrong – partly through acclimatisation (Gillbank 1986; Selleck 2001; Minard 2013). We must also take into consideration the relatively recent science of 'biogeography' pioneered by Alexander von Humboldt (Jeffries 1997) and the broader international thinking of 'environmental renovation' (not restoration!) promulgated by proto-conservationists, such as the American George Perkins Marsh (Tyrrell 2004) and French acclimatiser Isidore Geoffroy Saint-Hilaire, who were big influences on Australian acclimatisers. Apart from contextualising the introduction of species like carp to Australia in the 19th century, this understanding may also make us a little humbler concerning more recent conservation, scientific and cultural approaches to global environmental and social issues, such as 're-wilding' and food security.

But let us begin with the environmental context into which the acclimatisation societies were founded and which have such a bearing on how the common carp fits into the bigger picture of colonial Victoria.

THE GOLD RUSH AND ITS ENVIRONMENTAL DEVASTATION

It is hard to overstate the influence of the gold rush on every facet of the fledgling colony of Victoria in the 1850s. The population of Victoria increased from a mere 70,000 in 1850 to more than 500,000 less than 10 years later (Frost 2013). This created greater demand for all sorts of goods: food, clothes, digging equipment, carts, horses, fodder, tents, lanterns, fuel, tobacco, tea, sugar, grog and everything else that goes with mining and camping out in the bush. Gold miners arrived in ever-increasing numbers from around the world, especially Britain, California and China (Younger 1970; Hughes 1996), closely followed by entrepreneurs who made money by meeting the diggers' needs. Many of the more prominent entrepreneurs were enterprising

Fig. 3.1. Gold mining, Chinese encampment, Guildford near Castlemaine, 1861. Source: R. Daintree, public domain, State Library of Victoria.

expatriate Americans, who started a range of businesses that proved enormously successful and lasted well after the heady days of the gold rush were over (Younger 1970; Humphries 2023).

For all the wealth it brought to Victoria, gold mining devastated both the land and the freshwater ecosystems with which it was associated (Fig. 3.1) (McGowan 2001; Frost 2013). Apart from the erosion caused by the carts of thousands of diggers making their way through the bush to and from the diggings, miners routinely set fire to the bush to prepare the ground: partly to get rid of snakes, and partly to expose gold that might be lying on the surface. A natural bushfire in February 1851 helped clear the vegetation and made surface gold more visible. But three years later, William Howitt noted that wherever the diggers went in the Buckland Valley, burning followed close behind (Howitt 1858). Fire was not useful just for clearing the bush to find gold. It also kept diggers warm at night, cooked their meals and was simply an enjoyable pastime – a way to rebel for many young men who had until then lived in cloistered and restricted communities back home.

The large population of hungry miners needed enormous quantities of food to keep their strength up during their long and arduous labours (Humphries 2023). Land under agriculture increased dramatically in the decade following the start of the gold rush in 1851, which meant that the bush was further cleared for growing crops such as wheat, and for market gardens

(Frost 2013). The animals that survived fire and clearing were still not safe from the diggers. As mining was prohibited on Sundays, bored miners often spent the day hunting native wildlife with their dogs. Everything was fair game: kangaroos, possums, native birds, goannas, snakes. The loss of wildlife was enormous (Frost 2013).

Not surprisingly, digging was a major cause of environmental and Aboriginal heritage site (Lawrence *et al.* 2018) degradation in the gold fields. It changed the landscape in ways that we can still see 170 years later. Among other effects, it allowed native weedy species to encroach, while extirpating less-tolerant others. Because this book is primarily interested in fish, to us the most significant fact is that creeks and streams were directly impacted by gold mining, as well as indirectly through the extraction of gold from the earth. Early on, creeks were dammed and the dry beds worked over for gold. What followed – the processing of the mined earth – can only be termed catastrophic.

Gold mining used enormous quantities of water. Every last drop of available water was used for sluicing (unless it fortuitously rained or a well was dug). Not only were streams pumped dry, but water that made its way back into streams was 'thick and foul with gold washing' (Howitt 1858, p. 110). The effects of this sludge were so severe – destroying market gardens, causing increased flooding in towns, fouling stock water, killing orchard trees – that newspaper articles discussed it relentlessly. Eventually, parliamentary inquiries, including a Royal Commission in 1859 and a Select Committee in 1861, investigated the sludge problem (Lawrence and Davies 2014; Davies and Lawrence 2019). The sludge affected almost three-quarters of all catchments in Victoria and its legacy is still evident today, although until recently it received comparatively little attention (Lawrence and Davies 2014; Davies *et al.* 2018; Davies and Lawrence 2019; Rutherfurd *et al.* 2020). The rivers and creeks around gold mining areas were rapidly and severely polluted, with undoubtedly disastrous effects on native fish and other organisms, although these were mostly documented anecdotally at the time.

> *We know that into the Moorabool the sludge sluggishly finds its way, killing the fish, and injuring the water* (The Colonial Mining Journal, Railway and Share Gazette and Illustrated Record, *4 October 1860, p. 23).*

Considering the scale of the sludge (and the fact it contained heavy metals and other contaminants), its effects would have dwarfed those of sediment input into streams by recent Victorian bushfires, which are much better studied and are known to have caused widespread fish kills and have long-lasting impacts (Lyon and O'Connor 2008; Silva *et al.* 2020).

A dense population, widespread indiscriminate burning, sludge in phenomenal amounts and the shooting of wildlife willy-nilly all combined to create an environmental tragedy of epic proportions (Frost 2013). This devastation did not go unnoticed by the general public and it certainly caught the attention of a phalanx of government scientists, medical practitioners, engineers and members of parliament; an inevitable outgrowth of the expanding Victorian economy fuelled by the very same gold (Hoare 1966; Minard 2015). The booming economy and the critical mass of population intensified the need for governmental organisations to regulate and legislate. They also catalysed the founding of universities, museums, libraries, art galleries and other scientific organisations and societies, including acclimatisation societies (Hoare 1966). Typically led by well-heeled scholars, often recruited from the UK or Europe, they embodied preconceptions and ideas that (in many cases) were based on the best science of the time, but in the end would prove arguably as disastrous as the problems they were trying to solve.

THE ORIGINS OF ACCLIMATISATION

Linden Gillbank noted that most staple fruit, vegetables, cereals and sources of meat eaten by the average person each day in Britain in the 19th century, as well as the animals that carried them around and pulled their ploughs, were not native to the British Isles (Gillbank 2001). Some, like cattle and wheat, were probably brought in as far back as the neolithic (Fairbairn 2000; Colledge and Conolly 2007; Smith *et al.* 2015; Cummings and Morris 2022), others came in with the Vikings, Romans or Normans and still others were more recent. Wheat, barley, oats and rye were introduced from the Middle East; millet, rice and pulses from Asia; and maize, beans and squashes from the Americas (Gillbank 1986; Anderson 1992). Indeed, humans throughout history have moved animals and plants around to diversify their diets, increase productivity and spice up an otherwise dull dish (Vitousek *et al.* 1997; Crosby 2000; Secord 2016; Elton 2020). In colonial times, this traffic went both to and from colonies. This is not intended to excuse colonial importations of organisms that later became feral. Rather, it is to note that transporting familiar animals and plants when moving to new lands or islands is common human behaviour. Certainly doing this on an industrial scale – 'invasion', it is sometimes called – makes the practice much more significant and thus more devastating to native vegetation, wildlife and the indigenous peoples of the lands to which the new species are brought. In the end, however, like in Europe and England, introduced species become 'naturalised' and are accepted as part of the country's fauna and flora. But this takes time.

During their colonial periods, most European powers established botanical gardens in their home countries, often government-sponsored ones that the public could visit to enjoy exotic animals and plants from overseas, which initially had more utilitarian purposes (Anderson 1992). Gardens were also often established at the residences of the nobility or royalty, primarily to impress dignitaries and other important guests (Pfennigwerth 2013). Kew Gardens in London began collecting plants brought back from naval expeditions in 1759, with the aim of promoting crops (e.g. sugar, tea, cinchona) which could if successful, be grown locally and in other colonies. Many colonial powers also established botanical gardens in their tropical or sub-tropical territories, where they experimented with growing local and exotic plants. The French, notably in Algeria, experimented with attempts to replace local crops with more profitable ones. Although many of the projects failed, such attempts were the beginnings of what would become the highly popular practice of acclimatisation.

Acclimatisation began as the French *acclimatation*, first used in 1776 by the enlightenment French writer G.T.F. Raynal as *acclimater* and later evolving into *acclimatement* (the action of familiarisation or adaptation of oneself with something new) by Louis-Sébastien Mercier, playwright and natural philosopher (Anderson 1992). In 1820, the term turned up in English as *acclimatation* and *acclimation*. All these terms were passive, relating to the manner in which animals and plants are affected by the climate in which they live. By the 1830s in France, the term had transformed into *acclimatisation*, which implied a more deliberate act of transforming species through their introduction to foreign countries. A decade later *acclimatisation* had made its way to England. *Acclimatisation* for the French was a logical development that stretched back at least to Swedish biologist Carl Linnaeus and France's Georges-Louis Leclerc, Comte de Buffon, who believed that God had created animals in a particular location and associated with a particular climate. It was consistent with Jean-Baptiste Lamarck's influential theory of the inheritance of acquired characteristics – the first coherent mechanism offered for how evolution operates – that intrinsically linked organisms to their climate and environment.

In France, acclimatisation was led by scientists working with and in museums, zoological and botanical gardens. It was deeply scientific and ostensibly practical (Anderson 1992). 'In France acclimatisation meant the investigation of the continuous

Fig. 3.2. Isidore Geoffroy Saint-Hilaire, 1855–70. Source: Franck, Wellcome Collection gallery CC BY 4.0.

process of transformation that living organisms undergo in alien environments' (Anderson 1992, p. 136). Its counterpart in Britain was led by gentlemen breeders and 'entailed the assembling of diverse, resilient creatures that, when alive, might entertain, and, when dead, might prove instructive to students of comparative anatomy' (Anderson 1992, p. 137). The French were interested in natural history. The British wanted ornamental garden plants and animals they could shoot.

The Société Zoologique d'Acclimatation was founded in Paris in 1854, with Isidore Geoffroy Saint-Hilaire (Fig. 3.2) as inaugural president (and Professor of Zoology at the Muséum d'Histoire Naturelle and director of its menagerie) (Gillbank 1986). Its manifesto was:

> to people our fields, our forests, and our rivers with new guests; to increase and
> vary our alimentary resources, and to create other economical or additional
> products (Buckland 1861, p. 7).

It was soon followed by regional associations, and in 1858, 33 acres (13.4 ha) in the Bois de Boulogne, Paris were allocated for its Zoologique d'Acclimatation. Membership of the Société

Zoologique d'Acclimatation soon grew to 2,000. Many of the nobility, and even some royals, were among its ranks.

An English naturalist and writer, Francis Buckland, founded the Society for the Acclimatisation of Animals, Birds, Fishes, Insects and Vegetables within the United Kingdom in London in 1860 (Fig. 3.3). His focus was animals rather than plants, because the latter were well catered-for by Kew Gardens (Gillbank 1986). Buckland had been much influenced by Saint-Hilaire in France and, as it happened, by Edward Wilson, editor of *The Argus* in Melbourne, who later founded the Acclimatisation Society of Victoria. Wilson happened to be in England

Fig. 3.3. Left to right, top to bottom: Francis Buckland, Edward Wilson, Ferdinand von Mueller, Frederick McCoy and Alfred Selwyn. Sources: FB: Wellcome collection gallery; EW: Maclure & Macdonald, 1848–78 National Library of Australia; FvM: J.W. Lindt, State Library of Victoria; FMc: Johnstone, O'Shannessy & Co., 1891, State Library of Victoria; GS: G. Moretti, Bibliothèque National de France. All CC or public domain.

around the time the UK society was formed – he was seeking treatment for eye issues – and he became its vice president in 1865. The UK society's fortunes waxed and waned. It struggled to garner as much support as its French counterpart, and amalgamated with the Ornithological Society in 1866. Buckland's interests lay mostly with fish; after 1867 he obtained (and retained for life) the post of Government Inspector of Salmon Fisheries, and later was instrumental in transporting salmon and trout eggs between England and Australia and New Zealand (Osborne 2000).

Acclimatisation in Victoria

In Australia, Edward Wilson had been an advocate of bringing in English songbirds, introducing alpacas and moving Murray cod outside its natural range (Gillbank 1986). He was not particularly impressed by the practical nature of Australian fauna and thought that 'every country should be effectively tested with every natural production for which it may be believed in any way suitable' (*The Times London*, 20 October 1858, p. 9). *The Argus* in Melbourne in the meantime argued that British people had relied on the introduction of fruit, vegetables and animals since Roman times, and that Australian nature (and agricultural technology) needed a similar boost, but over a much shorter period (*The Argus*, 29 October 1855, p. 4).

By the end of the 1850s, the Victorian government and local and London-based newspapers were all for diversifying what until then had been a reliance on gold and wool (Gillbank 2001). The proposal to introduce exotic animals and plants that would increase agricultural, horticultural and aquacultural production received widespread support (Gillbank 1986). Apart from the need to increase food production, when gold became increasingly scarce there were large numbers of people looking for productive work (and leisure), and thus a need for land for them to farm. Land, until then dominated by wealthy – and politically well-connected – squatters would have to be made available to former diggers ('yeoman' farmers: Powell 1989) to grow crops and raise animals for themselves and for profit. For production to successfully feed the growing population, it was especially important that the plants and animals were familiar to the farmers and were hardy enough to withstand the terrible conditions left by gold mining. One obvious solution was to introduce plants and animals from the old world that people knew how to handle, were tough, grew fast and were good eating.

The issue of the environmental legacy of the gold rush, not least the freshwater environment, which we described at length above, was uppermost in the minds of journalists, scientists and politicians – and proto-acclimatisers. There was a perception that the native freshwater fish had not been particularly abundant or diverse even before gold mining wiped out the few that did swim in Victoria's freshwaters (Minard 2015):

> *What a dreary land would this be to good old Izaak Walton! The race of anglers*
> *is in great danger of dying out in these colonies from sheer want to occupation.*
> *In the very best of times – the pastoral period of our history, before Mammon*
> *[gold] turned every river and creek into sludge channels – fish were anything but*
> *plentiful and the varieties were of a limited description (*Yeoman and Australian
> Acclimatiser, *5 September 1863, p. 37).*

Francis Buckland read a paper to the Society of the Arts of London in November 1860, and it was republished by the Acclimatisation Society of Victoria in 1861:

> *It is a happy provision of nature that there is a fish to be found adapted to almost any sort of water, from the lordly salmon of the mountain torrent to the humble eel of the stagnant ditch. Why should not we pay a little more attention to the habits of fish, and transfer fish suited for a certain kind of water in that water, supposing there to be none there already? Let us study the transport of fish and utilise waters, whether great or small, which are now idle (Buckland 1861, pp. 25–26).*

Buckland was referring to the UK, and not just introducing fish to other regions but culturing them as well. Australian acclimatisers republished Buckland's address because they believed his ideas made sense and were as applicable to Australia as they were to the UK:

> *But we must not flatter ourselves that we are altogether about to introduce pisciculture as a novelty into this country.*
> *In former days, when the inhabitants of this country were for the most part Roman Catholics, and, therefore, greater eaters of fish, the cultivation of fish was looked after; and I would quote a good authority on this point:–*
> *'That carp were introduced from the Continent to England by the monks is nearly certain ... While England was Catholic, great attention was paid the raising and fattening of the choicest varieties of fresh-water fish, an art which sunk into neglect, partly owing, doubtless, to the abolition of fast days, and partly to the great facility with which the finest sea-fish are now transported throughout the country.'*
> *There is not reason why we should be behind our ancestors in this matter, therefore let us set to work and see what is to be done (Buckland 1861, p. 29).*

Besides the gold rush, and largely as a result of the increased population and wealth that followed it, in the 1850s a flurry of scientific government positions in Victoria were created by governors La Trobe, Hotham and Barkly (Hoare 1966). Victoria was quickly leaving New South Wales behind scientifically, achieving more in a decade than the latter had in almost 50 years. Sound scientific advice was essential for a growing colony, not least when it related to gold (Gillbank 1986). Appointments included Alfred Richard Cecil Selwyn as Director of the Geological Survey in 1851; Ferdinand von Mueller as Government Botanist in 1853 and Director of the Botanic Gardens in 1857; and Frederick McCoy as foundation Professor of Natural Science at the University of Melbourne in 1855, Government Palaeontologist with the Geological Survey in 1856 and Honorary Director of the Museum of Natural and Applied Sciences in 1857 (Fig. 3.3). All three men were later prominent members of the Acclimatisation Society of Victoria (ASV).

Several societies preceded the ASV, and they shared many of the same members. In 1854, the Victorian Institute for the Advancement of Science and the Philosophical Society of Victoria were founded. A year later, they amalgamated into the Philosophical Institute of Victoria, which became the Royal Society of Victoria in 1860. The Zoological Society of Victoria was formed in 1858. Its objectives were many and the last one included specific mention of acclimatisation:

> *Fifthly: It is contemplated that a portion of the gardens be set apart and arranged*
> *for the care and protection of birds, mammalia etc., which private individuals*
> *may import into the colony, in order that by particular care and attention they*
> *may become* acclimatised [our emphasis], *at such terms as may be agreeable to*
> *the discretion of the Society (*The Argus, *1 March 1858, p. 8).*

The Argus on 9 March 1858 sang the new Zoological Society's praises:

> *The Zoological Society may confer a lasting boon upon the country by*
> *establishing the several sorts of deer, antelopes, and gazelles, which would swarm*
> *in the unenclosed country, unmolested by beasts of prey ... and, by help of tanks in*
> *communication with the Yarra, the various kinds of fresh water fish might be bred*
> *and conveyed by degrees into all the rivers and fresh-water lakes of the colony*
> *(*The Argus, *9 March 1858, p. 4).*

The Zoological Society was in many ways a precursor to the ASV, which took on many of its objectives. Prior to the founding of the ASV, there was already strong support for introducing useful and ornamental animals from overseas into Victoria. The Zoological Society menagerie soon held an impressive list of native and exotic mammals and birds, including emus, koalas, Sumatra deer, Californian quail and English pheasants, which the public enjoyed immensely (Gillbank 1986). The Victorian government was particularly enthusiastic about any attempt to bring potentially productive plants and animals into the colony.

The ASV was founded by Edward Wilson in 1861 (he had returned to Australia in 1860, with alpacas and songbirds in tow), the same year as the founding of the New South Wales Acclimatisation Society. Founding members included von Mueller, McCoy and Selwyn – and notably Thomas Embling MD and MP. Wilson, as we have already seen, had been instrumental in setting up the equivalent UK society and, through his editorship of *The Argus* and his involvement in the Philosophical Institute of Victoria, had for several years advocated the tweaking of 'God's plan' (Wilson 1857). He was able to convince most of the Zoological Society to join his new organisation, whose objectives were virtually identical to the British version – a focus on acclimatisation, bringing in exotic plants and animals and spreading them around the colony.

The ASV soon took over the property of the Zoological Gardens Management Committee, which was transferred to the new ASV-run Zoological Gardens at Royal Park. Initially a place to store animals after their arrival from overseas, the public interest in viewing the exotic beasts led to the gardens soon becoming the Melbourne Zoo.

> *The ASV had arisen on a wave of enthusiasm for animal and plant*
> *acclimatisation. With both public and government support, it tried to please*
> *everyone. For the pastoralist it offered the camel, alpaca, angora and cashmere*
> *goat; for the sportsman, deer, elk, hare, quail, and various types of duck; for the*
> *angler, salmon, trout, carp and other fish; for the agriculturalist, the (supposedly)*
> *grub-eating birds such as the thrush, blackbird, starling, sparrow, and Indian*
> *mino (minah); and for the cottager, the Ligurian bee (Gillbank 1986, p. 372).*

DARWIN, FISH NAMES AND ENVIRONMENTAL RENOVATION

To complete our discussion of acclimatisation and especially the ASV, we need to emphasise three additional issues that put into historical context the enthusiasm – or at least low opposition – for acclimatisation worldwide. The first was antipathy to Charles Darwin's and Alfred Russel Wallace's theory of evolution through natural selection; the second was the state of animal taxonomy and classification at that time; and the third was ideas about conservation through 'environmental renovation'.

First, McCoy, von Mueller and the British and French acclimatisers Francis Buckland and Isidore Geoffroy Saint-Hilaire, from whom they sought inspiration, were all anti-Darwinists (Osborne 2000; Minard 2013). All believed that God had 'placed' animals and plants at sites of creation. There was no adaptation through natural selection to a particular environment, such as in Australia. Humans could tweak God's work by shifting organisms around, if the climate was conducive to their survival and productivity. That was, according to the anti-Darwinists, fitting and allowable because of the role entrusted to Man by God (Osborne 2000). McCoy, especially, considered successful acclimatisation as evidence that evolution was hogwash. How could an introduced species do better than a native species, if adaptation to the environment was how things worked? His ideas were very influential at the ASV; that is, until successful acclimatised animals, like sparrows and mice, became major farm pests. McCoy saw this success as positive evidence for his own theories. Others were less convinced.

Second, the taxonomy and classification of plants and animals in the 19th century was a far cry from what it is today (Minard 2015). For example, many of the Australian freshwater fishes were considered to be in the same families as those in the Northern Hemisphere; hence the common names that included the words *cod*, *perch*, *minnow* and *trout* for species that were completely unrelated to their counterparts in 'the old country' (Humphries and Walker 2013). Because of this misunderstanding and confusion, people were less concerned about introducing animals to Australia: they saw it as bringing in other members of the same family, but ones that were known to be hardier than their native cousins. This was especially the case if the species were brought into waters that had been severely damaged by gold mining or other degrading influences. Indeed, Edward Wilson, in seeking species of fish to acclimatise in Victoria, was advised not to import salmon but instead to bring in trout, carp and tench, because they were considered hardier species (Minard 2015). Wilson was not convinced, and continued to advocate for salmon and alpacas as ways of improving food supply and agricultural production.

Finally, although influenced by French and British acclimatisation, Australian acclimatisation was part of a global phenomenon that included the US (Minard 2015). Von Mueller and Wilson especially, along with others in the ASV and the broader community, were heavily influenced by the ideas of American George Perkins Marsh which were encapsulated in his hugely popular book *Man and Nature* first published in 1864 (Fig. 3.4).

Members of the ASV were also influenced by the father of biogeography, Alexander von Humboldt, who expressed concerns about the destruction of forests (Tyrrell 2004). Marsh was one of America's first 'conservationists' and he was especially keen on forests and fish. His philosophy was that humans had had profound effects on nature and climate, and that through judicious use of aquaculture, in the case of fish, much of that damage could be undone. Philosophically and practically, however, Marsh was not a pure conservationist but an 'environmental renovator' – someone who wanted to improve degraded systems through the introduction of tolerant species (Tyrrell 2004). He recognised that there was no going back to Eden, but there was an opportunity to shape the future:

> *... the artificial breeding of domestic fish has already produced very valuable results, and is apparently destined to occupy an extremely conspicuous place in the history of man's efforts to compensate his prodigal waste of the gifts of nature. The restoration of the primitive abundance of salt and fresh water fish, is one of the greatest material benefits that, with our physical resources, governments can hope to confer upon their subjects. The rivers, lakes, and seacoasts once restocked, and protected by law from exhaustion by taking fish at improper seasons, by destructive methods, and in extravagant quantities, would continue indefinitely to furnish a very large supply of most healthful food, which, unlike all domestic and agricultural products, would spontaneously renew itself and cost nothing but the taking (Marsh 1864, p. 118).*

Along with many of the members of the ASV, Marsh recognised that the Victorian bush needed to be fixed after the catastrophe caused by gold mining. He also recognised that

Fig. 3.4. George Perkins Marsh, 1855–65 and *Man and Nature*, 1864. Source: Library of Congress, public domain.

Australia had the wealth and scientific firepower to get something useful done. There was a moral dimension to this philosophy too, which was taken up by *The Argus* and *The Age*:

> *In a new country such as this, one can see to what extent man is a levelling agent, and how easily he can disturb the harmonies of nature. Mining, ploughing, roadmaking, the cutting of drains, the formation of tracks, all aid in diminishing the conservative powers of the natural heritage, and it is not surprising that our best streams, such as the Loddon, Campaspe, and Avoca, are fast becoming mere channels for the efflux of sludge and sand ... The reservation of large tracts of forest land is our first duty. By keeping the hills clothed we may make fruitful the valleys, and provide moisture for the parched plains* (The Argus, *16 October 1865, p. 5*).

*****

Nineteenth-century Australia, especially Victoria, was ready, willing and able to import animals and plants seen as productive, edible and able to survive the rigours of the Australian climate and the legacy of destructive gold mining. Although many introduced species soon became pests, the early introductions of carp were apparently relatively harmless. Indeed, it was not until an introduction almost 100 years after the first one that things really got going. In the next chapter, we explore the various introductions of carp, before getting to the main event – when carp went from a relative rarity in our rivers and lakes, to become the dominant species.

4

The introduction of carp to Australia

It is stated in Paris, Berlin, and Hamburg the carp is preferred to every other fish except trout and salmon (Southern Argus, *10 June 1880).*

THE STORY OF THE INTRODUCTION OF THE COMMON CARP INTO AUSTRALIA is complicated because multiple introductions occurred on and off for 100 years from as early as 1858. Further, the introductions were made by numerous groups – mostly acclimatisation societies – at numerous locations in various colonies. It is also complicated because most introductions were not successful in establishing populations, as the fish did not survive or did not breed. There were no rigorous fish surveys done at the time, so any information we do have is anecdotal. Most of all, it is complicated by the uncertainty of which species of 'carp' was introduced during any one particular event. As we move through time, however, the story gets much simpler. But there is no comfort from this simplicity. Instead, it reveals the tragedy of the introduction of the Boolarra carp into the Murray River.

Carp was not the only fish introduced during the heady days of acclimatisation in mid-19th century colonial Australia (Fig. 4.1). At least 43 freshwater fish species have been brought into Australia in the last 150 years or so (Harris 2013). Like carp, the introduction of trout and salmon (Family Salmonidae) to Australia was a foregone conclusion, given their popularity in Europe. The salmons and trouts were and still are considered by Europeans, and by people in many other regions of the world, angling royalty (Clements 1988). English, European, Eurasian or redfin perch (*Perca fluviatilis*) was another species high on the list of imports (Weatherley 1974; Harris 2013). A good angling species, it copes with a broad range of water temperatures, habitats and water quality. It also grows fast and tastes good. It was one of the first alien fish species to be introduced (1858 in Tasmania and 1861 in Victoria: Clements 1988; Harris 2013; Lintermans 2023), and indeed it is not uncommon to meet anglers who think that redfin is native to Australia.

Other cyprinids, including roach in 1861 and tench in 1876, were introduced because of their angling and/or eating qualities (Harris 2013). But the top contender for the most frequently introduced of all introduced fish to Australia has to be the goldfish, often referred to historically as golden carp. Starting in the early 1860s and continuing to this day, it has been introduced for

Fig. 4.1. Common alien freshwater fishes introduced to Australia in the 19th century. (A) European or redfin perch. (B) Roach. (C) Tench. (D) Goldfish. (E) Crucian carp. Sources in order: Gilles San Martin CC by SA 2.5; Karelj, public domain; Karelj, public domain; US Fish and Wildlife Services, public domain; Karelj, public domain.

its aesthetic qualities as an ornamental fish for ponds and fountains. When released into the wild and when it breeds, however, the goldfish reverts to its ancestral type, loses its bright colour and can grow to a considerable size (Lintermans 2023). Large size and drab colouration, among other things, add to the uncertainty of which species of 'carp' was introduced at a particular time and location. As mentioned in the Introduction, common carp and goldfish can hybridise (with progeny having one pair of whiskers or barbels reduced sometimes to buds), which adds to the uncertainty.

Another species of carp worth mentioning is the crucian carp, a more cold-hardy species than the common carp. Many authors over the years have written about the supposed introduction of crucian carp to Australia (e.g. Lake 1967; Weatherley and Lake 1967; Wharton 1971) but it seems that crucian carp and goldfish are easily confused, and most fish that were in the past called the former, were in fact the latter (Clements 1988). Indeed, there were no crucian carp in any Australian museum collection checked by Doug Hume in the early 1980s, and all fish named as crucian carp were in fact goldfish (Hume *et al.* 1981). Extensive fish surveys have turned up a few individuals of the species in the Campaspe catchment of Victoria since 2004 (Lintermans 2023). So there must have been an introduction of crucian carp at some time in the past, but individuals are few and far between. There was supposedly a fourth carp introduced, the Gibel carp (*Carassius gibelio*), which was sometimes called the Prussian carp. But there is no solid evidence of this species having been introduced, and Clements thought that the Prussian carp, to which ichthyologists like Gilbert Whitley referred (Whitley 1951), was in fact the common carp (Clements 1988).

This chapter documents the many early introductions of common carp into Australia, pinpointing as much as possible where the introductions took place, how many fish were introduced and who was involved. We aim to give a measure of certainty about whether it was the common carp that was introduced in the early days. But doubt can never be entirely eliminated without a description, illustration or photograph of the actual fish – and those are rare. From Shearer and Mulley's pioneering genetic study in the late 1970s (Shearer and Mulley 1978) to more recent work by Davis and colleagues (Davis *et al.* 1999) and Haynes and colleagues (Haynes *et al.* 2009), most scientists agree that there are three dominant existing 'strains' of common carp in the Murray–Darling Basin: 'Prospect', 'Yanco' and 'Boolarra'. It appears that more recent multiple introductions of 'Koi' have also taken place. Genetic analyses tell us that the strains have interbred, which could be an important explanation as to why the common carp has exploded since the Boolarra fish got into the Murray in the 1960s. But for now, we follow the many introductions of carp to Australia that began 100 years before Boolarra.

Common names and the common carp

Common names can be confusing and were used inconsistently in the past, especially in newspaper reports. We use 'common carp' or 'carp' as the common names for *Cyprinus carpio*, the focus of this book, 'goldfish' for *Carassius auratus* and 'crucian carp' for *Carassius carassius*. We follow Clements (1988) when assessing historical records of introductions of species. He concluded that past ichthyologists, like David Stead, Gilbert Whitley and John Lake, may have at times made poor choices for common names but nevertheless could identify the true species to which they referred. We and Clements are not convinced that this was the case with acclimatisers, however, and so we note uncertainty about which species is referred to if 'carp' is simply used (we give these 'low probability' of being *C. carpio*), except when 'goldfish' or 'golden carp' are used in the same statement to distinguish between them and the true common carp. Having said that, goldfish were valued by Europeans almost exclusively as an ornamental fish. They were kept in ponds to entertain and soothe with their bright colours, not for angling or for eating. So, it seems logical to conclude that if 'carp' were released or 'liberated' into large water bodies, like lakes and especially rivers, the species released would most likely have been the common carp. We allocate these 'carp' a 'moderate probability' of being *C. carpio*. We further assume that European carp, English carp, English river carp,

> great carp, King carp, common carp, Prussian and German carp are in all likelihood *C. carpio*, and allocate them a 'high probability'. This is mostly because the scientific name has often, but not always, been given alongside these common names. Because of these assumptions, our assessment of the number of introductions of *C. carpio* may be slightly inflated but that does not change the fact that there have been many separate introductions to Australia of the common carp.

SINGING CARP'S PRAISES

From the late 1850s, there was a flood of enthusiasm for importing the common carp to regions of Australia that might favour its survival, growth and breeding. Some of this, as we have already seen, came from the Acclimatisation Society of Victoria (ASV). It is true that carp was not in the same league as trout and salmon in the eye of the acclimatisers, and, with other species, collectively termed 'specimens of the not very valuable pond fishes of England' (Acclimatisation Society of Victoria 1864, p. 32). But not everyone had this low opinion of carp. Newspaper articles across Australia compared the carp's flavour favourably with those of angling royalty:

> *... the carp will thrive in almost any ditch or puddle, so long as it is moderately*
> *clear, and has some green stuff growing in it. In Germany the breeders plough*
> *up a low piece of ground, sow rye upon it, and when it is well up turn on water*
> *together with carp, and in a short time the carp get quite large and fat. It is stated*
> *in Paris, Berlin, and Hamburg the carp is preferred to every other fish except trout*
> *and salmon (*Southern Argus, *10 June 1880).*

Indeed, although one writer had never eaten a carp, he concluded that 'from their universal celebrity as a pond fish throughout England, no doubt they are toothsome and good' and would 'be worth trying in the Hunter and its tributaries' (*Maitland Mercury and Hunter River General Advertiser*, 27 February 1883, p. 2).

People had mixed feelings about its angling qualities (remember, it was called the river 'fox' by English anglers because of its wily nature in avoiding capture). Carp was at times lauded for its potential for fish farming (e.g. *The Mercury*, 10 November 1883, p. 1) and at other times for the sport it provided for fishers (e.g. *Mount Alexander Mail*, 15 October 1872, p. 2). As we saw in Chapters 1 and 2, carp had few rivals that could match its combination of fecundity, growth rate, broad diet, hardiness and ability to be transported easily: 'they can be kept alive for weeks in cool places, if wrapped in moss and fed bread soaked in milk' (*Southern Argus*, 10 June 1880, p. 3).

Introductions of common carp to the US were often cited as examples of the advantages of importing this particular species (in the 1880s, carp were not yet a pest in the US). Specifically, comparisons were made between California and Australia because of their similar climates.

> *... it is found [in America] that carp thrive in warm and shallow waters, and it*
> *is therefore more than probable they would be easily acclimatised in Queensland.*

> *Notwithstanding the fact that carp are considered rather coarse as an addition to*
> *our present scanty fish supply, they are well worth the trial, and their introduction*
> *would be a powerful inducement to land-owners to make a water supply which*
> *would enable them to enjoy the luxury of a good fishpond (*The Queenslander,
> *28 July 1883, p. 31).*

Local and broader distribution of carp was also advocated by members of the general public with European connections:

> *... a German friend of mine ... is of opinion that the European carp would do*
> *exceedingly well in the Torrens Lake, as well as in the lagoons on the Murray*
> *and the lakes in the South-East, and would much like to see the Acclimatization*
> *Society take the matter up (*South Australian Register, *letter to the editor, 25*
> *August 1883, p. 7).*

Carp was even seen as having potential to eradicate the freshwater snails that were intermediate hosts for sheep live fluke (*The Australasian*, 7 November 1903, p. 53).

While some of the journalists and letter writers in the last years of the 19th century were aware of acclimatisers' attempts by to import the common carp, others were less informed. The first introduction of carp had in fact taken place many years before.

EARLY INTRODUCTIONS OF COMMON CARP TO AUSTRALIA

The best contender for the first common carp to reach Australian shores (still alive after the long sea voyage) is a solitary individual imported into Tasmania in 1858 at the behest of Morton Allport (Figs 4.2 and 4.3) (Clements 1988). Allport was a naturalist, lawyer and driver for the introduction of salmon and trout to Tasmania. He was brought to Hobart Town by his parents

master Burchall.

The *Heather Bell*, Captain Harmsworth, arrived on the 22nd February, after a remarkably fine passage of 89 days. She spoke no vessels connected with the colony on the passage. The *Heather Bell* brings the first lot of live English fish to the colony. Out of about 120 placed on board, Captain Harmsworth has succeeded in landing 6 in all, consisting of tench, perch, and carp. They are imported by Mr. Moreton Allport, who hopes to be successful in stocking the waters of the colony, to which the imported fish will be a valuable addition. Every voyage Captain Harmsworth makes some addition to our stock of useful and fancy birds; and he deserves the warmest thanks for the lively interest he takes in landing them in good condition. The *Heather Bell* has been but little over six months from this port, and the homeward voyage from Melbourne was performed in a remarkably short time of 69 days.—*H. T. Courier.*

Fig. 4.2. Morton Allport and shipping news from *The Argus* (2 March 1858, p. 4), concerning the importation of carp on the *Heather Bell* on 22 February 1858. Source: Allport Library and Museum of Fine Arts, State Library of Tasmania: Morton Allport Portrait Photo, Hobart, Tasmania (photographer unknown), SD_ILS:603598.

in 1831 while still a baby (Stillwell 1969). Brought up immersed in natural history and art, he became an expert in the botany and zoology of Tasmania, corresponded with scientists in Europe and became one of the first salmon commissioners in 1866. In addition to the carp, he brought tench and European perch to Tasmania. The fish were released into a reservoir at Cascade near Hobart Town (Clements 1988). Apparently the introduction was successful. It was reported in *The Mercury* on 12 April 1862 that the tench were doing very well after some had been placed in the Hobart Botanic Gardens (then distributed further). The carp had bred and there were, after four years, 70 swimming around in the Cascades Reservoir. In 1869, Allport claimed that he could supply 'brown trout, perch, tench and carp' to 'those who can guarantee that proper care will be taken to carry out the experiment with success' (*The Tasmanian Times*, 29 April 1869, p. 2; originally from the *Launceston Times*), which suggests he had enough carp available to dispatch to interested parties. There is some uncertainty, however, concerning the species of carp imported. R.M. Johnston, writing almost 30 years later, mentioned 'The Crucian Carp ... *Carassius vulgaris*' and 'The Golden Carp ... *Carassius auratus*' and stated that 'These are to be found in our various rivers, and are so well known that they need no description' (Johnston 1882, pp. 64–65). We beg to differ.

The next introduction of the common carp was probably in 1860. This is based on a letter to the editor of *The Argus* by Edward Wilson from London, which was published on 12 January of that year. Much of Wilson's letter talked about how and where to best introduce salmon, the thrushes and blackbirds he had brought out to Australia and how Queen Victoria herself took a keen interest in acclimatisation and offered any help she could with the activities:

> *Thus, then, I may say literally, that from the highest to the lowest – from the Queen herself to the humblest fisherman on the banks of the river – the colonists have received every indication of sympathy and co-operation (*The Argus, *12 January 1860, p. 5).*

Wilson went on to say:

> *... will the time be early or remote in which their country [Australia] may abound with every good thing that the earth produces in any way suitable to the country. You [the Australian people?] have got the alpaca in fair process of trial. You are getting the camel. Sir George Grey sends you some of the most valuable venison animals of South Africa. Captain McMickan has given you the carp (*The Argus, *12 January 1860, p. 5).*

Later that year, in another letter to the editor of *The Argus*, Wilson stated that 'I have also sent out by Messrs. Wigram's last ship, another consignment of carp and perch' and expressed his wish that an aquarium built on board the ship *Lincolnshire* would routinely bring fish to and from England over the coming years (*The Argus*, 19 December 1860, p. 5).

In 1861, Wilson listed the animals held in the Melbourne Botanic Gardens as including 'a quantity of carp, tench, dace, roach, and goldfish' (*The Argus*, 8 August 1861, p. 5). In April 1862 a further consignment of five carp and seven goldfish (plus six grey mullet, one eel, two French newts, tench, some birds and dogs) arrived by the *Lincolnshire* (*The Argus*, 28 April

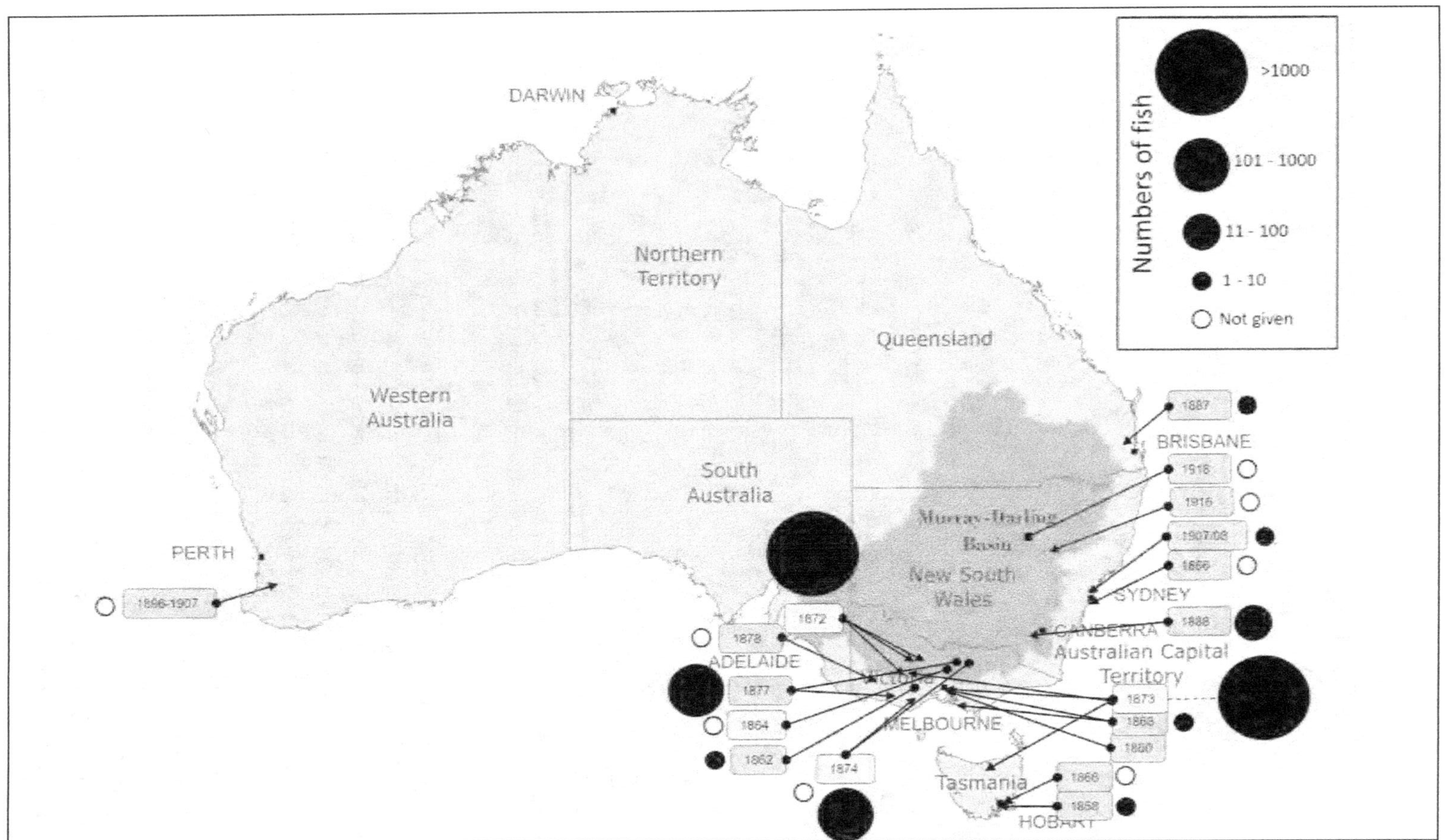

Fig. 4.3. Dates and locations of pre-1960 introductions of carp to Australia. Greyscale relates to the certainty of the introduction being common carp, *Cyprinus carpio*. Black = high certainty; white = possible, but uncertain. Circle size indicates categories of numbers of fish introduced. For details of introductions, including numbers of fish, exact location and sources of data, see Appendix 1.

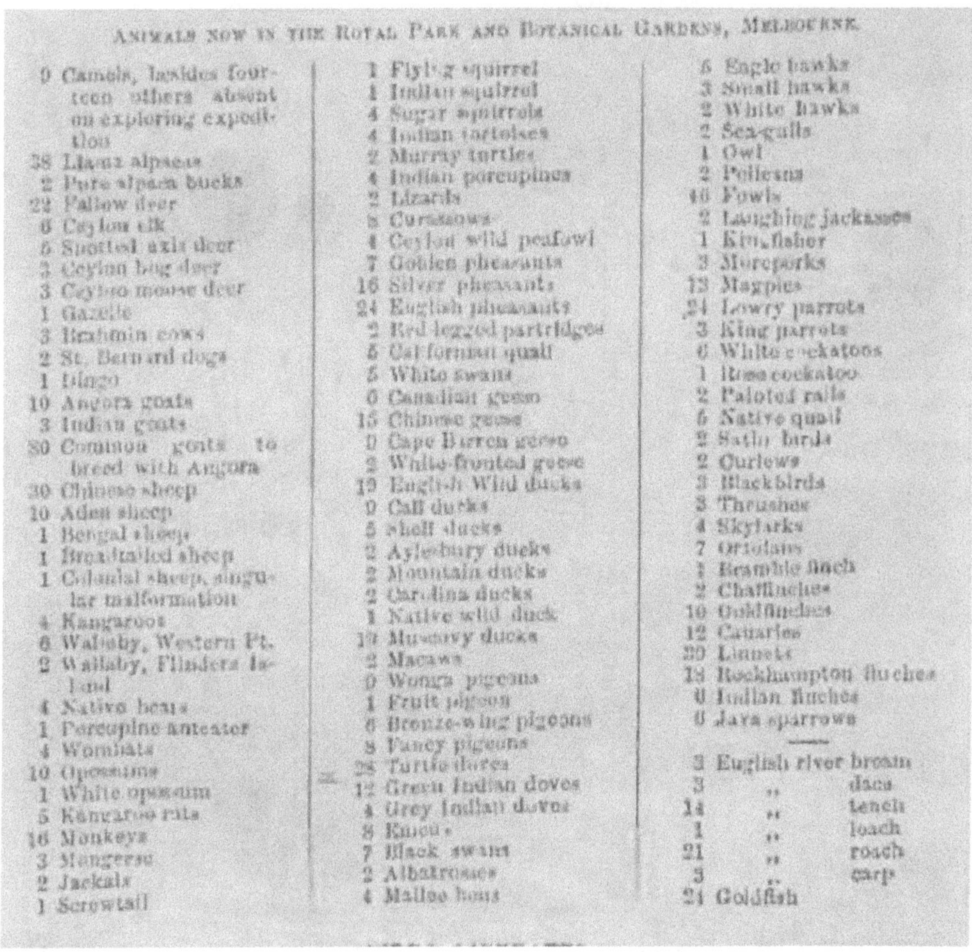

Fig. 4.4. Inventory of the animals being kept in the Melbourne Botanic Gardens and Royal Park. Note: the fish at the bottom include three English river carp and 24 goldfish. Source: Acclimatisation Society of Victoria 1862, p. 42.

1862), probably destined for the Botanic Gardens. These newspaper records, and the ASV 1862 Annual Report (Fig. 4.4, bottom right), confirmed that in all likelihood the fish from multiple shipments were released into ponds at the Melbourne Botanic Gardens.

We can be fairly certain that the fish kept in the Botanic Gardens were common carp, because Clements' parents remembered seeing these red, orange and yellow carp during the 1920s and 1930s. Clements saw them in the 1940s as a child and helped destroy hundreds of them, some of which weighed 15–25 lb (6.8–11.3 kg), as part of the carp eradication program in 1962 after the species was declared noxious (Clements 1988; Chapter 5). The photographs in his article show conclusively that the carp destroyed were common carp.

During the 1860s, there were several further introductions of carp by the ASV (Acclimatisation Society of Victoria 1862, 1865, 1866; McCoy 1862), some of which were released at 'various locations favourable to their multiplication' (Acclimatisation Society of Victoria 1864, p. 32).

They were only referred to as 'carp', so their identification is uncertain. Other introductions were mentioned by the Acclimatisation Society of New South Wales (ASNSW). These fish were called Prussian carp, but they were referred to separately from golden carp:

> *The society are gratified that the golden carp has been very recently naturalised in New South Wales, together with another species (the Prussian carp), by the exertions of his Excellency Sir John Young, who has bred them in ponds near Government House, and has distributed several of the progeny to congenial situations elsewhere, so that we may soon expect them to be abundant in the colony ... Several of these fish given to this society, and placed in the tank in the Botanic Gardens as Prussian carp, eventually turned out golden carp, and it is now ascertained that the carp (probably both species) distributed by his Excellency are swarming in some of the lagoons and creeks between Sydney and Botany. They were very recently captured in great numbers, several being from four to six inches in length* (Sydney Morning Herald, *26 April 1866, p. 2)*

A year later, the fifth annual report of the ASNSW stated that 'The golden and Prussian carp has [sic] been naturalised' (*Tasmanian Morning Herald*, 7 May 1866, p. 3).

There was a hiatus of a few years, followed by another flurry of introductions in the 1870s (Fig. 4.3), most of which were reported only as 'carp'. Although some fish were released into ponds or lakes, others were released into the Wimmera River, Victoria (1872) and or 'liberated in the bush' in large numbers (Zoological and Acclimatisation Society of Victoria 1872–1878, 1873, p. 35; 1874, p. 11; 1875, p. 12). While we will never know for sure, the fact that fish were released into rivers suggests that they were more likely common carp than goldfish. Other acclimatisation societies had also got into the action. The Blackwood Fish Acclimatisation Society west of Melbourne introduced 'European carp' into Blackwood Reservoir, Bacchus Marsh probably in 1874:

> *... we have much pleasure in bringing under your notice the success of the Society in acclimatising the European Carp in the waters of the Reservoir. That they have thriven and multiplied was exemplified last summer, when seven carp from 3 to 6 inches in length were captured a short distance below the reservoir* (Bacchus Marsh Express, *4 December 1875, p. 3).*

The Geelong and Western District Fish Acclimatising Society reported in 1875 that it had been presented with several 'carp', some of which were placed in ponds and others in 'waters under the control of Mr Raddenbury' (*Geelong Advertiser*, 1 September 1875, p. 3). A newspaper report two years later stated that the *SS Derwent* had arrived from Launceston with 'about 1,000 carp and tench, for the Geelong Fish Acclimatisation Society' (*The Australasian*, 10 March 1877, p. 19). A report in 1881 indicated that the carp 'in the local streams were increasing rapidly' (*Australasian Sketcher with Pen and Pencil*, 18 June 1881, p. 203).

By the end of the 1870s, newspaper articles and the annual reports had confirmed the introduction of common carp to Tasmania, Melbourne and Bacchus Marsh, and the likely 'liberation' of the species in central and western Victoria and in various 'creeks and lagoons' around Sydney. Some of the Victorian 'liberations' involved literally thousands of fish (Fig. 4.3). Although in many instances the true species imported and released will forever remain a mystery, the presence of common carp in later commercial fishery catches (Chapter 7) and the recent genetic work may clear up some of the uncertainties.

There were few credible carp introductions between the 1870s and the early 1900s (Fig. 4.3). But there was the odd mention in isolated newspaper articles. One example was an article in the Brisbane-based *Telegraph*, noting that a Mr D. O'Connor visited Hobart and Ballarat and brought back 47 perch, 10 trout and two Prussian carp for release into Enoggera Reservoir. Apart from a few fish that died on the way, the rest apparently were healthy (*The Telegraph*, 21 June 1887, p. 4).

Notably, however, between 1896 and 1907 the Western Australian Acclimatisation Committee brought in many exotic freshwater species including eels, Atlantic salmon, European perch, tench and carp (Fig. 4.3). An article from 1910 stated that only perch and brown trout had established populations:

> *In regard to fish acclimatisation, much had been done without a great deal to show of it (Here the lecturer showed some fine pictures in connection with perch fishing in the Collie river.) With perch they had been very successful, though the same could not be said of the tench and carp and Murray cod (*Bunbury Herald, *20 December 1910, p. 3).*

However, a piece published the year before in the Bunbury *Southern Times* stated that 'Mr Toombes, the local Postmaster caught in a crab net, some small English perch, Prussian carp and Blackfish' (*Southern Times*, 6 November 1909, p. 4), which suggests that carp had survived for at least a couple of years after release.

A carp introduction worth mentioning, before we attempt to unravel the most significant introductions that pre-date the Boolarra saga, is a single record of carp being transferred to Lake Windamere, near Mudgee, north-west of Sydney. We discuss this because it made it into the 1916 New South Wales Fisheries Department Report and so has some authority: 'A successful transplant of Prussian Carp to Lake Windermere [sic] was conducted early in the year' (Reid *et al.* 1997). Forty-eight fish up to half a pound [0.25kg] in weight were released, reportedly sourced from the Prospect ponds near Sydney. There are no follow-up mentions of this population, nor stories in any newspaper that we could find. The 'Prussian Carp' were most likely *C. carpio*, if we follow Stead's and Clements' advice on common name usage.

PROSPECT STRAIN

One of the most enduring populations of common carp in Australia was introduced via the Sydney aquarium trade and at the hand of one of Australia's leading ichthyologists, David George Stead. We know all about it, because he told us the details in an article he wrote for the *Australian Zoologist* some years later (Stead 1929). There had been some speculation about the origins of those particular fish. Being a very competent ichthyologist, Stead unequivocally

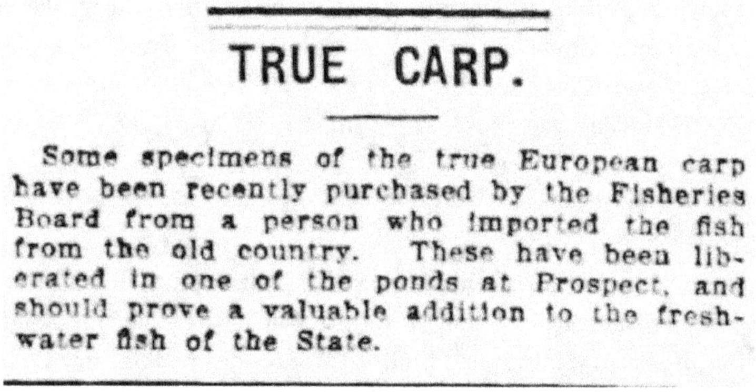

TRUE CARP.

Some specimens of the true European carp
have been recently purchased by the Fisheries
Board from a person who imported the fish
from the old country.   These have been liberated in one of the ponds at Prospect, and
should prove a valuable addition to the freshwater fish of the State.

Fig. 4.5. Newspaper article mentioning the introduction of common carp to the Prospect Reservoir, Sydney. Source: *Sydney Evening News*, 19 December 1907, p. 5.

confirmed the identification of the species he introduced, and shed light on the loose use of common names for the species:

> *One day as I was examining the fish stocks of a bird and animal dealer at a shop in George Street, Sydney, right opposite Bridge Street, I noticed among several hundreds of plain and golden carp* Carassius auratus, *of from two to four inches in length, a few which possessed the typical barbels of* Cyprinus carpio. *A close inspection of these revealed that they were indeed of the species named, and which is known variously as 'Asiatic', 'Prussian', 'German' and 'Common' Carp, in literature [Incongruously, Stead referred to the fish in the title of his article as 'The Great Carp* Cyprinus carpio']. *Twelve of these, ranging from two to two and a half inches in length, were secured – at a cost, it may be mentioned of sixpence each. This was December 11, 1907 (Stead 1929, p. 100).*

In fact, goldfish and 'Prussian carp' had been sold by Sydney merchants since at least 1868, 40 years earlier! Advertisements for aquaria, 'fish globes', goldfish and 'Prussian carp' appeared in several newspapers during that period (*Sydney Mail*, 26 December 1868, p. 13). Of the fish purchased by Stead, he released nine into Bloxsome Pond, an inlet pond for Prospect Reservoir which supplied water to Sydney. Three of them he kept in a fish tank. The event was reported in the newspapers, confirming Stead's date and version of events (Fig. 4.5). It has since been shown the 'Prospect' strain of carp started by Stead were imported from Europe (Haynes *et al.* 2010).

A year later, Stead bought a further six common carp, placed them in the raceway of the trout hatchery at Prospect Reservoir and hand-fed them for about 18 months. By 1917, there were 93 carp in Prospect Ponds and a year later 49 were recorded (Clements 1988). Gilbert Whitley, doyen of all things ichthyological and an expert historian, stated that some of the fish were later 'transferred to other enclosed waters, whilst other fish from the same 1908 batch are still living in Taronga Park Aquarium' (Whitley 1951, p. 234). While both Stead and Whitley mention the confusion about which carp species was introduced when and where, Stead, like Clements, was convinced that Whitley confused his common names but not the actual species. Whitley stated that the crucian carp was released into the Queanbeyan River in 1888 'and is now a pest inland' (Whitley 1951, p. 234). They were most likely goldfish that had lost their colouring and grown large. But he may have been correct.

YANCO STRAIN

The existence of a 'Yanco' strain of common carp has been confirmed as distinct from those Stead released into Prospect Reservoir by all authors who have studied the genetics of carp in the Murray–Darling Basin (Shearer and Mulley 1978; Mulley and Shearer 1980; Davis *et al.* 1999; Haynes *et al.* 2009). Mind you, interbreeding and hybridising with goldfish has muddied the water somewhat. Yanco carp most likely formed a self-sustaining population before the Prospect and Boolarra strains (see Chapter 5) began spreading (Haynes *et al.* 2009). Where did those particular fish come from?

A New South Wales Fisheries Report of 1931 mentioned the revival of the Aquarium Society of New South Wales and the importation of tens of thousands of fish from Asia, some of which were koi from Singapore (Clements 1988). But the statement is a long way short of evidence that someone connected with, or who purchased fish from, the Aquarium Society transplanted some of those carp into the Murrumbidgee Irrigation Area (MIA) (Shearer and Mulley 1978). Tantalisingly, Clements also referred to a conclusion by Shearer in a 1977 article that carp were released into the Murrumbidgee prior to 1903, most likely from Indonesia. Unfortunately, this article is not in Clements' reference list and we have not been able to find it. But with a mystery like this, like all good researchers, we decided to investigate further.

We have spent considerable time trawling through digitised newspapers in the Australian National Library's Trove repository, starting from 1931 and working backwards in time. And we may have found the smoking gun for the origins of the Yanco strain of common carp! The following is how we think the Yanco carp was introduced to the MIA and we provide the evidence – via newspaper articles – that we hope you will find convincing. We will let you be the judge.

Although carp were first introduced to Hobart Town in 1858, Charles Macarthur of Launceston, Tasmania introduced 'English carp' (*The Age*, 21 December 1865, p. 7), presumably to northern Tasmania at least by 1865 (Fig. 4.6). The English carp most likely came from Hobart, possibly from Morton Allport, since he was offering services in that regard more widely a few years later. Wherever they originated, pisciculture was clearly taking place in northern Tasmania in the late 1800s. For example, in 1876 a consignment of English carp was sent from Launceston, via the steamer *Derwent*, to the Geelong and Western District Fish Acclimatising Society (*Geelong Advertiser*, 16 February 1876, p. 2).

In 1870, the Ballarat Fish Acclimatisation Society (BFAS) was formed. It initially focused on trout for angling in Lake Burrumbeet, but soon expanded to other species of fish and other lakes and streams (Fig. 4.7).

In 1872, the Tasmanian fish culturists/acclimatisers, continuing their association with the mainland, sent thousands of trout eggs over Bass Strait to the Zoological and Acclimatisation

Society of Victoria, as it was known at that time, the Ballarat Fish Acclimatisation Society and the Clunes and Castlemaine Acclimatisation Societies (Fig. 4.8).

That same year, the *Mount Alexander Mail* reported – in addition to trout culture – the release of 48 English carp in the Expedition Pass and Harcourt reservoirs near Castlemaine, expressing the hope that in a few years fishermen 'can enjoy a day's sport equal almost to any English fishing' (*Mount Alexander Mail*, 15 October 1872, p. 2). In 1873, the Castlemaine Acclimatisation Society (CAS; formed in 1872) received 214 small carp (*The Ballarat Star*, 28 August 1873, p. 4) from the Victorian Zoological and Acclimatisation Society. They were released into the Commissioner's Gulley Reservoir. European perch was also mentioned and indeed, perch

of fish. The council are glad to be able to inform the society that Mr J. A. Youl has, in conjunction with Mr Edward Wilson, again consented to undertake the arduous and important task of superintending the collection and shipment of this ova in London. Through the kindness of Mr Charles Macarthur, of Launceston, the council have also been enabled to introduce English carp. Whilst treating of fishes, the council would observe that they have taken every opportunity to aid in the development of the fisheries of this colony. Taking advantage of

Fig. 4.6. Newspaper article mentioning the introduction of 'English carp', possibly to northern Tasmania or Victoria, from the annual meeting of the ASV. Source: *The Age*, 21 December 1865, p. 7.

The Gipps Land people are now engaged in introducing brown trout from Tasmania into the rivers of that region. In Ballarat a society has been organised as "The Ballarat Fish Acclimatisation Society," with the object of introducing English trout to the waters of Lake Burrumbeet and other suitable reserves of the Ballarat district. A committee, consisting of Drs Whitcombe and King, and Messrs Rowlands, Seal, Finn, Cowan, Learmonth, and M'Ewan, is appointed to make preliminary arrangements for the introduction of the ova from Tasmania.

Fig. 4.7. Newspaper article introducing the formation of the Ballarat Fish Acclimatisation Society. Source: *The Ballarat Star*, 12 August 1870, p. 4.

ACCLIMATISATION.—The Tamar, which sailed yesterday afternoon at 4 p.m. took away four cases of brown trout ova, containing 2,000 for the Zoological and Acclimatisation Society, Melbourne; 2,000 for the Ballarat Fish Acclimatisation Society; 1,000 for the Clunes Acclimatisation Society; and 500 for the Castlemaine Acclimatisation Society.

Fig. 4.8. Newspaper article mentioning a shipment of trout eggs from Tasmania to Victoria. Source: *The Mercury*, 22 August 1872, p. 2.

usually preceded carp whenever introductions were made. Later that year, the BFAS was busy establishing breeding ponds in the Ballarat Botanical Gardens. It also received 84 3-inch carp from the CAS, although there was some uncertainty as to the actual species (some members of the BFAS thought that they were gourami [*Osphronemus goramy*], a native of south-east Asia which looks superficially like a carp).

In addition to distributing carp and perch to various waters, such as Lake Burrumbeet, Lake Wendouree in Ballarat was fast becoming the focus of stocking and angling (*The Ballarat Star*, 30 October 1873, p. 2). Carp and tench were directly acquired by the BFAS from a Tasmanian friend of Ballarat seed merchant R.U. Nicholls (*The Ballarat Star*, 8 December 1873, p. 2). *The Ballarat Star* gave the actual numbers of fish received in 1873. Eighty-three carp had arrived from the CAS (one had died); 48 of them were put into Lake Wendouree, 25 into Lake Burrumbeet and the rest in other lakes; 60 carp had come from Tasmania thanks to R.U. Nicholls and were all kept in the Ballarat Botanical Gardens ponds (*The Ballarat Star*, 22 April 1874, p. 3). We have not been able to find details of the origins of these carp, although they most likely came from breeding ponds in northern Tasmania (maybe at Clarendon), possibly through the Northern Fish and Game Protection Society. Members of that society had removed carp and/or carp eggs from the fountain in Prince's Square, Launceston in September 1873 'for the purpose of placing them in some of the rivers' (*The Tasmanian*, 20 September 1873, p. 5) and for 'breeding purposes' (*Cornwall Advertiser*, 19 September 1873, p. 2), so carp had clearly been in northern Tasmania for at least a few years by then.

In 1876, *The Ballarat Star* spruiked the English carp as a 'fine fish' that would do well if introduced to the nearby Lake Learmonth, claiming that, among other advantages, it would not harm other fishes in the same waters. The fish was, we are told, able to be acquired from Tasmania. The article distinguished it from other varieties of 'carp about this district being a smaller fish than the common English carp (*Cyprinus carpio*), and not half so valuable' (*The Ballarat Star*, 7 March 1876, p. 3). So, goldfish had likely been introduced to the area by this time as

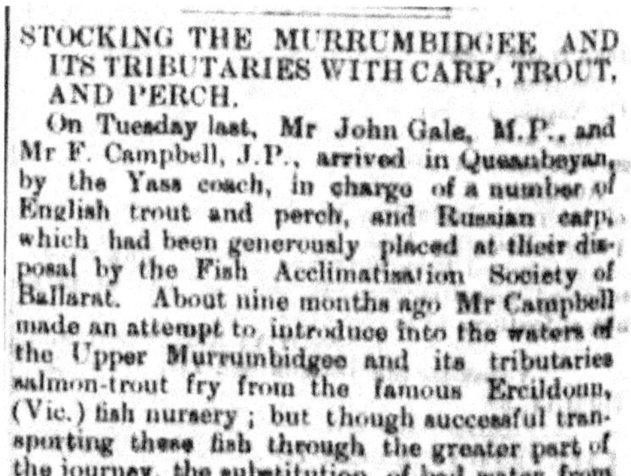

STOCKING THE MURRUMBIDGEE AND ITS TRIBUTARIES WITH CARP, TROUT, AND PERCH.

On Tuesday last, Mr John Gale, M.P., and Mr F. Campbell, J.P., arrived in Queanbeyan, by the Yass coach, in charge of a number of English trout and perch, and Russian carp, which had been generously placed at their disposal by the Fish Acclimatisation Society of Ballarat. About nine months ago Mr Campbell made an attempt to introduce into the waters of the Upper Murrumbidgee and its tributaries salmon-trout fry from the famous Ercildoun, (Vic.) fish nursery; but though successful transporting these fish through the greater part of the journey, the substitution of bad water from a tank at Junee poisoned the fish, and at Harden they were found to be all dead. The present attempt, has, however, been eminently successful. Messrs Gale and Campbell left the Ballarat station on Monday morning last, and by Tuesday evening the greater part of their living freight were in the Queanbeyan, Molonglo, Naas and Cotter Rivers (tributaries of the Murrumbidgee.) To give some idea of the care necessary to ensure the safe transport of the fish in charge of these gentlemen (especially the trout) we are informed that their piscatorial treasures were placed in tins specially constructed for the purpose on principle approved of by the Acclimatisation Society, and to ensure the agitation and aerating of the water necessary for the health of the fish a pair of bellows furnished with india-rubber tubing had to be employed at short intervals throughout the long and tedious journey. The railway authorities both of New South Wales and Victoria not only furnished free passes along their lines, but every other accommodation necessary for the safe transit of the fish. The journey by rail and coach occupied 36 hours, during the whole of

Fig. 4.9. Newspaper article concerning the transport of English trout, perch and 'Russian' carp from Ballarat to the Murrumbidgee. Source: *Mount Alexander Mail*, 7 July 1888, p. 2.

well. By 1878, carp were doing very well in Lake Wendouree, Ballarat. They had bred and were reaching nearly 3 lb (1.4 kg), apparently a boon for the angler (*The Australasian*, 28 September 1878, p. 13).

Now we get to the smoking gun. Carp had been 'quietly' going about its business for several years in Lake Wendouree, the BFAS ponds and elsewhere, breeding and being distributed hither and thither (*The Ballarat Star*, 17 July 1889, p. 2). Then, in 1888, it was reported in *The Argus* on 29 June, and in the *Mount Alexander Mail* and *The Ballarat Star* a week later, that John Gale MP and F. Campbell JP on 25 June had picked up a consignment of 300 small trout, perch and carp or 'Russian' carp from the BFAS and transported them by train, via Melbourne and Albury, to the upper Murrumbidgee catchment (Fig. 4.9). By the evening of 26 June 'the greater part of their living freight were in the Queanbeyan, Molongo [sic], Naas, and Cotter rivers, tributaries of the Murrumbidgee' (*The Ballarat Star*, 6 July 1888, p. 2). Carp may have been introduced to the Murrumbidgee catchment prior to 1888, but this was the earliest detailed account of carp being released into this watershed, which included the fish's origins and the identification of the agent.

There was a big flood in 1891 in the Murrumbidgee; one of the biggest on record. That same year, E. Morley of Gundagai caught what were thought to be 'English carp', which had 'only been noticed since the flood' (*Hay Standard and Advertiser*, 16 September 1891, p. 3). By 1908, a type of carp was caught even further downstream by Mrs Hill of Yanko Creek (Fig. 4.10). These fish were 'pronounced to be golden carp' and were considered a 'novelty' to the area (*Narrandera Argus and Riverina Advertiser*, 1 May 1908, p. 4). They may have indeed been goldfish. However, the Yanco strain of common carp is known to be brightly coloured and has small eyes. Fish are orange–red on the head and back and yellow underneath, with an orange–red caudal fin, and the Yanco/goldfish hybrid is greenish–gold in colouration (Shearer and Mulley 1978). Whether the fish that Mrs Hill caught in the Murrumbidgee near Bringagee were goldfish or common carp is of no great importance, because fisheries scientist H.K. Anderson confirmed the enormous numbers of 'common carp' in billabongs in the same area of the Murrumbidgee in 1917 (Anderson 1918). Thirty years later, fisheries scientist J.O. Langtry, carrying out a landmark survey throughout the lower Murray–Darling Basin, also caught what he referred to as 'carp' in 1949/50 at Bringagee on the Murrumbidgee (Cadwallader 1977).

Shearer and Mulley (1978) speculated, admittedly from a very limited sample set, that in 1976 Yanco carp were restricted to the MIA. Indeed, they defined the strain based on its location – all 34 fish were collected from the main irrigation canal at Yanco. They concluded that the strain was genetically distinct from the Prospect and Boolarra strains and had its origins in Asia. Subsequently, the

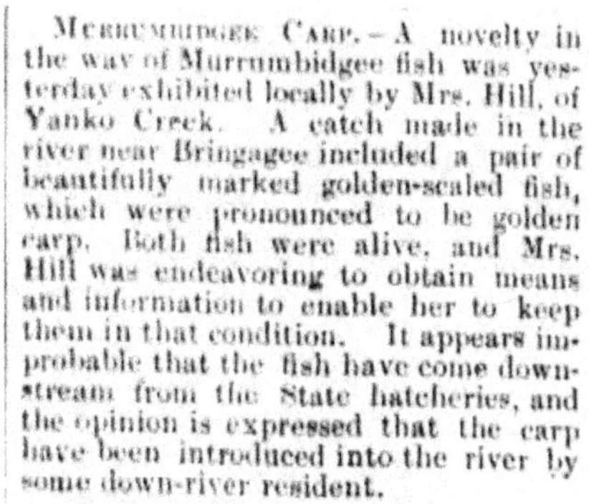

MERRUMBIDGEE CARP. – A novelty in the way of Murrumbidgee fish was yesterday exhibited locally by Mrs. Hill, of Yanko Creek. A catch made in the river near Bringagee included a pair of beautifully marked golden-scaled fish, which were pronounced to be golden carp. Both fish were alive, and Mrs. Hill was endeavoring to obtain means and information to enable her to keep them in that condition. It appears improbable that the fish have come down-stream from the State hatcheries, and the opinion is expressed that the carp have been introduced into the river by some down-river resident.

Fig. 4.10. Newspaper article with details of the occurrence of either Yanco common carp or goldfish in the Murrumbidgee River near Bringagee. Source: *Narrandera Argus and Riverina Advertiser*, 1 May 1908, p. 4.

strain was sometimes referred to as the 'Singapore' strain. Until now, who first introduced these fish to the MIA has been a mystery.

Clements (1988) disputed the restricted MIA distribution of common carp, claiming, as did others (Lake 1959), that carp were widespread throughout the Murray–Darling Basin but in low numbers prior to the Boolarra carp escape in the 1960s. Langtry's 1949–1950 survey (Cadwallader 1977) and commercial fisheries statistics from the 1940s and 1950s (Reid *et al.* 1997) (Chapter 7) offer support for that claim. There is more recent genetic support from fish collected in the mid-1990s. Davis *et al.* (1999) suggested that the Yanco strain, identified by Shearer and Mulley, was not only present in the MIA but had, at least by then if not before, dispersed as far as the Darling River. A decade later, Haynes *et al.* (2009) confirmed that interbreeding among strains of carp and dispersal of progeny meant that Yanco genes, although still dominant in the MIA, were present in carp throughout the Darling River and many of its tributaries. Koi genes distinct from Yanco genes were present in the wild in Australia, but Haynes *et al.* (2009) only recorded them from a few populations in the upper reaches of some rivers closest to the Great Dividing Range, whereas they dominated Parramatta River carp (Haynes *et al.* 2010) outside the Murray–Darling Basin.

*****

Since 1858, when carp were first imported into Tasmania, there were without doubt many introductions of common carp to Australia. Indeed, Tasmania and Ballarat, the latter via the BFAS, seem to have played important roles in distributing carp to the south-east mainland of Australia. Although we have struggled to find out much about carp culture in Tasmania, it was apparently northern Tasmania – likely through the exploits of the Northern Fish and Game Protection Society – from whence came many of the carp that were dispersed, especially in Victoria. We agree with Clements that there is 'no doubt, whatsoever, carp have been in Victoria for a very long time' (Clements 1988, p. 333) and in New South Wales, prior to the Boolarra 'carptastrophe' of the 1960s and beyond. While some newspaper records of importations and introductions were detailed and the weight of evidence is hard to dismiss, in the next chapter we present some of the more credible evidence that the common carp existed in the wild prior to the Boolarra carp's escape into the Murray River. We then tell the story of the Boolarra Fish Farms, how the business came about and its role in establishing populations of the most infamous of all carp in Australia.

5

Boolarra: A 'carptastrophe'

*The story of European carp (*Cyprinus carpio*) in Victoria is one of great tragedy and frustration (Wharton, 1971).*

BECAUSE OF THE APPARENT EASE WITH WHICH NON-NATIVE ANIMALS HAVE been introduced to new regions of the world, it is tempting to assume that establishing populations in new environments is a piece of cake: just chuck them in and they will grow. Luckily, it is not that easy (Williamson and Fitter 1996). But with enough introduction events and enough animals, the odds that introductions will succeed become much greater. First, let us be clear what we are *not* talking about, and then we will proceed to what we *are* talking about in relation to introducing species. Importantly, when it comes to species introduction, context – the animal's intrinsic traits and the external environment into which the animal is introduced – is everything. We will explain this more fully below, and hopefully it will reveal the analogous conditions that existed in medieval Europe and 19th- and 20th-century Australia.

What we are *not* referring to here are domesticated animals – dogs, cats, horses, cows, sheep, goats and the other animals that have been associated with humans for thousands of years (Diamond 2002; Zeder 2012). Our focus in this chapter concerns animals that are essentially 'wild'. These are animals humans have deliberately (e.g. brown trout, sparrows, cane toads and gambusia) or inadvertently (e.g. rats, mice, zebra mussels and sea lampreys) released into a new environment, sometimes far from their native range. For 'wild' animals, successful establishment of populations following an introduction is by no means guaranteed.

Releasing animals into the wild (even if the 'wild' environment means a pond in the Melbourne Botanic Gardens or a farm dam) means that individuals effectively have to fend for themselves. Many 'wild' species, though, have a history of introductions in numerous regions of the world. Therefore, often they have been selected because of their hardiness, fast growth rates and other 'useful' qualities such as sport (hunting or angling) and/or because they taste good. Through the process of repeated introductions, acclimatisers and others interested in such things worked out which animals were worth persisting with and which were a waste of time, money and energy (Crosby 2000; Low 2001). Today we only see the species that won, and it is left to historians to illuminate those that lost in the introduction stakes.

We have seen that there were many early introductions to Australia of common carp and, because of myriad confusing references, goldfish. These species had been well-and-truly tested over the centuries in Europe and Asia for their hardiness and the desirable qualities that made them – and some others – ideal to import into Australia: carp for angling and eating; goldfish for colour and movement. That there were dozens of introductions of both species is beyond dispute. What is much less certain is how many resulted in established populations.

For a freshwater fish to survive, breed and establish a population from an introduction, we know it has to have appropriate life history (e.g. breeding time, number of young, growth rate, age at maturity) and other favourable traits (e.g. broad diet, wide temperature tolerance). Other critical factors are the numbers of individuals involved and the environment into which the fish are introduced (Britton 2023). When carp were imported in low numbers into Australia, populations may not have established unless the environment was ideal. A pond or town reservoir, perhaps? No doubt some introductions of carp were unsuccessful (Koehn *et al.* 2000; Koehn 2004) but we know that at least four survived: Prospect, Yanco, Boolarra and Koi (Davis *et al.* 1999; Haynes *et al.* 2009; Haynes *et al.* 2010).

Until the Boolarra strain got into the Murray River and spread (what we call here the 'Boolarra event horizon' or BEH), the Prospect and Yanco strains were presumed to exist in relatively small areas: the immediate Sydney environs in the case of the former and the Murrumbidgee Irrigation Area in the case of the latter (Shearer and Mulley 1978; Brumley 1996; Koehn *et al.* 2000; Koehn 2004). Clements (1988), however, didn't go along with that presumption, and through researching newspapers and consulting scientific documents we lean more to his way of thinking. Recent genetic studies also seem to partially back up Clements' claims. It is likely that the Prospect strain may have been more widespread prior to the BEH than previously thought and could have been helped in its dispersal through the mega-floods of the 1950s, whereas the Yanco strain's spread may have been post the BEH (Haynes *et al.* 2009).

In this chapter, we briefly discuss evidence for the existence – and the possibility of established populations – of carp in the Murray–Darling Basin prior to the BEH. Then we shift gear and tell the story of the Boolarra Fish Farms. We describe how the combination of the intrinsic qualities of the carp and the sheer number and widespread nature of the introductions of Boolarra carp enabled the 'tragedy' referred to by Jim Wharton in the quote that starts this chapter. There were warnings by some scientists of the destruction the introduction of the Boolarra carp could wreak, and legislation was rushed through that made carp a noxious fish in Victoria: a species that was to be killed on sight. We describe how those with foresight fought to prevent the BEH from happening, and came within a barbel's breadth of success. But with carp – a super-spreader, a 'powerful invader', a fish that conquered Asia, Europe and the UK and that single-finnedly created what we have called the Carpocene – a barbel's breadth was the difference between calm and 'carptastrophe'.

CARP BEFORE BOOLARRA

What evidence is there that common carp populations existed, especially in the Murray–Darling Basin, prior to the BEH? Well, not a lot, but there is some. Again, some of the uncertainty relates to the common name confusion. We will attempt here to once again wade through the quagmire of confusion to the firmer ground of certainty.

In Table 5.1, we show sources that in our view are more reliable than newspaper articles, as they come from fisheries experts who would have been able to tell a common carp from a goldfish or crucian carp. Although journalists are not ichthyologists, the multiple lines of evidence presented in Chapter 4 for the introduction of what we conclude is likely to have been the Yanco strain of carp into the upper Murrumbidgee, and its spread downstream in the following decades, are nevertheless compelling. Our sources here draw on NSW Fisheries Department reports, published articles by fisheries scientists, commercial fish records and a few other sources from experts. Our results complement those of Clements (1988), who stated:

> *... the lack or sparsity of reports of carp catches by fishermen in those early days, particularly prior to 1960, does not mean they were not there and not being caught. Early commercial fisheries statistics, and government reports, only refer to 'carp' and prior to the 1950s freshwater fisheries research by qualified people was virtually non-existent and the small amount of research done by government departments was directed into the more valuable commercial native fisheries and the introduced trout. The introduced 'carps' and tench were such a low priority as to be, I suspect, almost completely ignored (Clements 1988, p. 335).*

To mention a few examples, beyond what we already know of around Sydney, H.K. Anderson and colleagues in 1917 'rescued' thousands of fish from drying billabongs along the Murrumbidgee River, including large numbers of small, probably juvenile carp (Table 5.1). Anderson, an experienced fisheries scientist, referred to the fish as 'common carp' and was detailed in his designation of native species, so we are reasonably confident in his identification that these were indeed *Cyprinus carpio*. Letters between Anderson and F. Lewis, Victorian Chief Inspector of Fisheries and Game, indicate the presence of common carp in Kangaroo Lake, Victoria, and in the Murray River, which is technically New South Wales. New South Wales Fisheries Reports from the early 1900s indicate carp or common carp in Lake George and around Corowa and Albury. And NSW Fisheries catch statistics from the late 1940s up to the BEH indicate that sometimes sizeable weights of common carp from multiple catchments and from rivers and lakes in the Murray–Darling Basin made their way to NSW fish markets (Pease and Grinberg 1995). It is clear, however, that from about 1971 commercial carp catches rose dramatically, peaking in the late 1970s. We will look at these in more detail in Chapter 7.

J.O. Langtry's fish survey in 1949/50 provided a lot of detail of catches over a broad area of the mid- and lower Murray and tributaries (Table 5.1). He collected small numbers of carp, presumably common carp, across a relatively broad area, as far west as Lake Victoria, almost to the border with South Australia and as far east as Yarrawonga. Also of note was that redfin perch and tench were even more widely distributed than carp, although numbers were still quite low.

There is little doubt, from these sources, that common carp was widely distributed across at least the southern Murray–Darling Basin prior to the BEH. We agree once again with Clements (1988), whose investigation, although pre-dating ours by almost 40 years and lacking access to all the material we have today, is remarkably comprehensive and insightful:

Table 5.1. Examples of evidence for the presence (not release) of common carp in Australian waters prior to 1960.

Source	Name	Location	Number/weight	Size
Reports of NSW Fisheries Dept, 1891	Carp	Lake George, NSW	'number of small carp increasing rapidly'	?
Stead (1929) mentioning 1907 and beyond	European carp, *Cyprinus carpio*	Prospect Reservoir and outlets, NSW	1917 = 93; 1918 = 49	
Anderson (1918)	Common carp	Murrumbidgee River near Bringagee, NSW	100s	Small
Reports of NSW Fisheries Dept, 1921	Carp or possibly English carp	Shallow waters near Corowa and Albury, NSW	'great number'	?
Letter from F. Lewis to H.K. Anderson, 1929	English carp, *Cyprinus carpio*	Kangaroo Lake, north-western Vic., on the Murray River	1	Adult
Letter from H.K. Anderson to F. Lewis, 1929	*Cyprinus carpio*	Murray River, Vic.	'odd ones'	?
NSW Fisheries catch statistics, Pease and Grinberg (1995)	Common carp	Many catchments and lakes throughout NSW	Highly variable catch weights from 1947/48	Market size
Langtry (1949–1950)	Carp`	NSW/Vic. from SA border to Yarrawonga	Lake Victoria (5), Frenchman's Creek (3), Rufus River (13), Lindsay River (24 kg), Boundary Bend, Murray River (1), Barmah Lake (some), Burramine billabongs (5), Yarrawonga (2), Murrumbidgee River at Bringagee (3)	?

Let there be no doubt, whatsoever, Carp have been in Victoria for a very long time. Although mainly recorded only from ornamental ponds in botanical gardens in Victoria, it was recorded in the Murray River system in New South Wales a number of times as early as the 1920s (possibly earlier) and although I have uncovered only one reliable record of Carp in northern Victoria (1929 at Kangaroo Lake) prior to the 1960s it would be fanciful not to think they were thinly, but widespread, in the Murray River drainage area of northern Victoria in those early days as they were in the same system on the New South Wales side of the border (Clements 1988, p. 333).

BOOLARRA

The origins of the fish farms

Boolarra is a sleepy little town in the Latrobe Valley, Gippsland, Victoria, not far from Morwell. It is home to about 1,000 people and is known for its annual folk festival, held in March, and its Australian Rules football team – the Boolarra Demons – that plays in the Mid-Gippsland Football League. Boolarra means 'plenty' in Gunaikurnai (Context 2019). The original inhabitants were the Brayakaulung clan of the Gunaikurnai and the Traditional Owners are the Gunaikurnai Land and Waters Aboriginal Corporation (https://gunaikurnai.org/our-story/), who have a 20,000-year connection with the area.

From colonial times, Boolarra was an agricultural and commercial centre supplying Melbourne and other markets with timber and farm produce. In the early 1900s, dairying took off and the population of Boolarra did likewise. Boolarra Fish Farms Pty Ltd was established in 1960 by Colin Eric Pratt and Franz Wucherpfennig, Mirboo North and Boolarra locals, respectively. The former wanted to encourage carp culture, and to distribute carp throughout Victoria to farmers so that they could get a feed from their dams. The latter was a German immigrant who survived the siege of Stalingrad during WWII and five years as a prisoner of war, and rekindled his passion for tropical fish and goldfish after emigrating to Australia in 1951 to work on the new Morwell briquette factory (*The Herald*, 16 April 1960). As we will see, Pratt's respect for government and scientific advice apparently were overridden by his eye for a lucrative business venture and his wish to keep that business afloat in the face of government opposition. The Boolarra Fish Farms eventually comprised 150+ ponds (Fig. 5.1) and were run by three generations of Wucherpfennigs. They bred and sold goldfish for almost 60 years, at one stage supplying 40% of Australia's goldfish market. As of 2024 they were winding down the business, mostly as a result of the trend towards importing goldfish from overseas rather than producing them in Australia. The once-thriving Boolarra Fish Farm ponds can still be seen from the air as a patchwork of rectangles.

To set the scene for the coming 'carptastrophe', we pause briefly and introduce you to some of the main players. Jim Wharton was manager of the Snobs Creek Fisheries Research Station from 1952 and deputy director of the Department of Fisheries and Wildlife from 1963 as the whole Boolarra carp saga unfolded. He wrote a detailed account of the lead-up to the importation of carp by Boolarra Fish Farms, and documented the attempts to stem the spread of

Fig. 5.1. Boolarra Fish Farms, showing the large number of existing and defunct ponds. Source: Map data ©2024 Google.

carp and eradicate it once it had been declared noxious (Wharton 1971). Alfred (Alf) Dunbavin Butcher, who had started as a biologist with the Victorian Department of Fisheries and Game in 1941, was director of the Department of Fisheries and Wildlife from 1960–1973 (Koehn *et al.* 2022) and became a strong opponent of the culture of carp and its trade with farmers in Victoria (Butcher 1962).

Jack Rhodes and his brother Phil were fisheries officers who witnessed first-hand the BEH and were part of the carp-kill team that poisoned more than 1,000 farm dams. They also killed the carp in the Melbourne Botanic Gardens ponds, that were descendants of fish imported almost 100 years previously (Rhodes 1999). Jack began with the Victorian Department of Fisheries and Game as an inspector in 1956 and worked in Wodonga from 1959 to 1977. He wrote a book on his experiences – *Heads and Tales: Recollections of a Fisheries and Wildlife Officer* (Rhodes 1999; Trueman 2011). While we know the background of Boolarra Fish Farms founder Franz Wurcherpfennig, that of Colin Eric Pratt is uncertain. We cannot find much reference to him or his activities beyond the carp controversy, and he seems to have vanished from the aquaculture scene not long after. Another player in our story is Geoffrey William Scott, general contractor, Boolarra. His role in the affair has, until now, not received much attention.

There were, of course, many energetic and committed people who attempted to stem the tide of carp. There were also some fisheries officers and members of the public who were not convinced that carp were a problem (Rhodes 1999). A few were pro-carp, but more were anti-poisoning. This was a time when there was heightened concern about the wholesale use of herbicides and pesticides – especially DDT and 2,4,5-T – exemplified by the publication of Rachel Carson's *Silent Spring* (Lear 1998). The presence, introduction and eradication of carp in Victoria in the early 1960s were controversial topics, much discussed in newspapers of the time.

The first hints of carp farming

In 1960, the Victorian Department of Fisheries and Wildlife received a request from Gippsland resident Colin Eric Pratt, who wanted to import 'European carp' 'specially bred in Germany' (Wharton 1971, p. 4). This was not the first sign that someone was interested in commercially farming carp. The Department had received letters as early as 1952 enquiring about the farming of common carp (Department of Fisheries and Wildlife 1952–1961). But Pratt's request rang alarm bells because of the scale of proposed operations. It was especially concerning because of reports from the US about the trouble carp had caused in waterways and to fauna and flora (Levi 1952; King and Hunt 1967). Pratt's enquiry was followed by several others from proponents of similar ventures, who wanted to provide an alternative to the importation of preserved fish, which they considered excessively priced (Department of Fisheries and Wildlife 1952–1961).

At an Inquiry into the Victorian Fishing Industry held in early 1960, Pratt argued that fish farming was the way of the future and would stabilise supply and prices:

> *Introductions of fresh fish farms in Victoria would increase the fish supply and help the consumer, the State Development Committee was told yesterday. Mr. C.E. Pratt, of Boolarra, near Morwell, told the committee that the industry could be established without any increase in existing resources, and utilise resources*

> *being wasted at present. Mr. Pratt said he and another Boolarra man, Mr. F.*
> *Wucherphennig, wanted to establish a fish farm. 'A fish farming industry can*
> *produce relatively unlimited supplies of fish of a required weight and at a required*
> *time'. Mr. Pratt said thousands of acres of water in Victoria were available to*
> *produce fresh fish. He said the committee which is examining the State's fishing*
> *industry, the Fisheries and Game Department, had refused an application to start*
> *a fish farm with carp. The reason given was that carp would take over Victoria's*
> *inland water system to the detriment of other fish. This was an ungrounded fear,*
> *Mr. Pratt said. Overseas research showed that the carp would not invade trout*
> *grounds as it preferred still water* (The Age, *12 May 1960, p. 12).*

It was also a way of using otherwise unproductive land. Native fish could not be farmed, trout and redfin perch did not grow fast enough and were temperamental when it came to aquaculture. 'European carp' was the fish of choice, and several varieties – mirror, leather, Aischgrunder, Galician, Lusation and Franconian carp – were available:

> *Colin Eric Pratt of Boolarra said Victoria's fresh fish supply could be greatly*
> *increased, and stabilised in both supply and prices, by means of fish farming.*
> *Thousands of acres, now unproductive, were available for the purpose. Native*
> *fish were unsuitable for farming. Of introduced species, only trout and redfin were*
> *available, and neither was suitable for large scale culture, being slow growing and*
> *difficult to control, and giving a poor yield per acre.*
> *The fish most widely used in pond farming was the Mirror or Line Carp*
> (Cyprinus rex Cyprinorum), *a sub-species of the European Carp* (Cyprinus
> carpio). *Several varieties of the sub-species were used: Leather, Carp,*
> *Aischgrunder, Galician, Lusation and Franconian being the main ones. The*
> *Leather Carp seemed to be most suitable. The scales had been bred off and the*
> *bones reduced in proportion to the flesh. There was no fish bred commercially*
> *that could equal the Mirror Carp for rapid growth. It could be marketed in*
> *weights up to 6–7 lb., and if market conditions were not favourable, it could be*
> *stored alive for relatively long periods* (Fisheries Newsletter, *July 1960, Vol.*
> *19, Issue 7).*

Pratt favoured the mirror carp, because it grew fast and attained a good size. He did not agree with the Department's claim that carp might be a problem if released into Victoria's waterways. After all, the species had been in Australia for a long time, had received no criticism nor done any 'apparent harm' and, as studies had shown overseas, was not a threat to trout (*Fisheries Newsletter*, July 1960, vol. 19:7, p. 13). Pratt used these arguments many times over the next year or two.

At the 1960 and 1961 Commonwealth–State Fisheries Conferences, the potential of carp introduction was not dismissed out of hand. Delegates were open to investigating the effects of carp on Victorian waters, flora and fauna. A committee set up by the Conference, however, decided that the importation of common carp should be prohibited for the time being. Pratt was told of this decision and, according to Wharton, apparently a 'gentleman's agreement' was made between him and the Department not to proceed with sales. But Pratt reneged. In July 1960, an advertisement appeared in the *Weekly Times*, avoiding the mention of carp but spruiking the benefits of raising fish in your own farm dam that you could eat within a year (Fig. 5.2). In his defence, Pratt later claimed that he had been forced to sell some carp to meet rising business costs including a delay in leasing a reservoir from the State Electricity Commission (Wiltshire *et al.* 1962b). Ironically, the Ballarat Fish Acclimatisation Society and the Victorian Piscatorial Council (an organisation representing anglers formed in 1901 that had once promoted the acclimatisation of fish) both expressed opposition to the introduction of common carp to Victorian waters (Department of Fisheries and Wildlife 1952–1961).

FISH

Open letter to all farmers,

Sir.

Would you like to have a plentiful supply of fat fish for your table? Fish that you can easily rear and fatten in your own ponds or dams? No doubt you would.

Our company hopes to be able to supply a fast growing and hardy fish suitable for stocking in a majority of dams and ponds. These fish, which would not require specialised handling or feeding, would be ready for the table within 12 months of being introduced into your waters, giving an assured supply of fresh fish at a very low cost.

If you feel there is a need for such a fish to be readily and cheaply available for stocking your waters please write to–

BOOLARRA FISH FARMS
BOOLARRA VICTORIA

Fig. 5.2. Advertisement in the *Weekly Times* encouraging farmers to grow fish (without mentioning that the fish was carp) in farm dams for the dinner table. Source: Redrawn from *Weekly Times*, 27 July 1960.

Carp are found at Boolarra

In August 1960, Victorian fisheries officer Phil Rhodes visited Boolarra Fish Farms and found several ponds specially built for the breeding and raising of carp. He also discovered some carp, including mirror carp, swimming in the ponds. There was speculation at the time that the fish had been imported from Germany, as Pratt had originally proposed, but Pratt later said that the fish were from the Prospect population in Sydney; a claim that he stuck to and Scott dismissed (Wiltshire *et al.* 1962b). More recent genetic studies and first-hand accounts suggest that Pratt was correct (Peter Rogan and Russel Wurcherpfennig, pers. comm.). The Department of Fisheries and Wildlife made an attempt through the courts to prevent Boolarra Fish Farms from stocking and breeding carp and then on-selling them to other farmers. Although importation into Australia was illegal, the magistrate ruled that since the waters were private, they were exempt (*The Canberra Times*, 21 March 1961). Pratt therefore won the case and was awarded 5 pounds and 5 shillings in costs, which the Department paid.

Boolarra Fish Farms resumed advertising the sale of carp (Fig. 5.3) and the Department subsequently discovered that young carp had been sold to farmers in various parts of Victoria to stock their dams. The advertisement used the German rather than the English word for carp – *karpfen* – and misspelled the scientific name as *Cyprynus* rather than *Cyprinus*. Pratt and co. were clearly not going to give up without a fight.

Following the court case in 1961, 'European carp' quickly became a hot topic in the public square. Many farmers and recent immigrants from Europe were pro carp. Others, anglers included, saw carp as the devil incarnate, especially because of its potential to affect their beloved trout.

Alf Butcher, director of the Department, took a trip to the US in August 1961 to see first-hand what carp were doing to the environment there, and sought advice from American fish experts on the carp issue. The directors of several US government fisheries agencies gave clear and vehement advice: do not introduce carp to places where it did not already occur (Butcher 1962). In the meantime, Pratt was far from idle. Although he later denied it, he apparently wrote a series of abusive letters to Arthur Rylah (*The Bulletin*, 10 February 1962), then Deputy Premier, Chief Secretary and Government Leader in the Legislative Assembly of Henry Bolte's Liberal Government (Costar 2002). According to the article in *The Bulletin*, Pratt had 'declared war' on the government because of its opposition to his carp scheme.

FISH

BOOLARRA FISH FARMS have pleasure in announcing that supplies of food fish for stocking farm dams and other private water are now available for immediate sale.

BOOLARRA KARPFEN

(Cyprynus Carpio)

can be obtained at the hatcheries at Boolarra. These fish are approximately 6 inches long and guaranteed to be of pure, healthy stock. They are fast growing, and at the correct stocking rates will be ready for the table within 12 months of being released

N.S.W. Fisheries Research Bulletin No. 5 classifies the eating qualities of the Cyprynus Carpio as "GOOD", and the angling qualities as "FAIR".

PRICE – £20 per hundred

As yet no rail or air deliveries can be made, but we will package fish in oxygen filled plastic bags and boxes for delivery at the hatchery.

Orders of 1000 or above will be delivered anywhere in Victoria or the Riverina at cost by the Fish Farms' transport.

For full particulars of stocking rates and methods of management write to

BOOLARRA FISH FARMS PTY. LTD.

BOOLARRA, VICTORIA

Fig. 5.3. Advertising carp by name, but in German, and spelling the scientific name incorrectly. Source: Redrawn from *Weekly Times*, 22 March 1961.

THE STATE GOVERNMENT INQUIRY

The controversy surrounding carp, the potential impacts of its release and the antipathy of government scientists towards the fish catalysed the Victorian Government to set up a State Development Committee Inquiry into the Introduction of European Carp into Victorian Waters, with Raymond John Wiltshire (Liberal and Country Party) as Chairman, George Roy Schintler (Labor) as Vice-Chairman and Frank Raymond Kenny as Secretary. Other members of the committee were the Honourables Buckley Machin (Labor), Arthur Robert Mansell (Country Party), Nathaniel Barclay DCM (Country Party) and Harold Edward Kane (Liberal and Country Party). The Inquiry ran for four weeks from 1 February to 1 March 1962 and took evidence from nine people. The Committee visited the ponds at the Boolarra Fish Farms and a reservoir

Table 5.2. Terms of reference, Inquiry into the Introduction of European Carp into Victorian Waters, March 1962 (Wiltshire *et al.* 1962a).

1. The question of whether the European Carp and its domesticated forms adversely affect the welfare of other species of freshwater fish and, if so, in what way?

2. Whether the European Carp and its domesticated forms adversely affect aquatic wildlife and, if so, in what way?

3. If the answer to question 1 or 2 is in the affirmative, should the propagation of European Carp and its domesticated forms be discouraged in view of the threat to established resources?

4. If the propagation of European carp and its domesticated forms is to be discouraged, is the problem sufficiently serious to require complete prohibition of propagation and, if not, under what conditions would propagation be permitted?

5. If it accepted that European Carp and its domesticated forms present a threat to established resources, should all European Carp and its domesticated forms, whether occurring in private waters or in waters defined under the Fisheries Act, be destroyed and, if so, by what means?

6. If the answer to question 5 is in the affirmative, what legislative actions is needed to permit this to be done?

7. Is the commercial propagation of the European Carp and its domesticated forms desirable in Victoria?

8. If European Carp and its domesticated forms should become readily available, is there reason for believing that there will be a steady and continuing demand for this species as a food fish?

9. If the European Carp and its domesticated forms should escape into natural waterways, would it be possible to remove it?

10. If legislative action is proposed to meet the problem presented by European Carp and its domesticated forms, is it desirable that such legislation should be sufficiently broad to provide for the control of and, if necessary, the extermination of, fish species which may become a problem at any time?

11. Any other matters which appear to the Committee to be relevant to the inquiry.

at Morwell, in which Pratt's company had placed juvenile carp. Members also 'examined and handled specimens of European Carp, including one of the original breeders brought into Victoria from New South Wales' (Wiltshire *et al.* 1962a, p. 8).

The Inquiry's terms of reference indicate a real attempt to determine whether carp posed a threat to Victorian freshwaters and their inhabitants. They also considered the destruction of all existing known carp, and addressed whether their destruction was feasible if they escaped their pond or dam confines and got into streams (Table 5.2). The Inquiry also considered if there was a demand for carp as food in Victoria, as Pratt had earlier claimed.

The subsequent report summarised what was known at the time about the taxonomy, morphology, biology, origins and history of introduction to Australia of the European or common carp, *Cyprinus carpio*. It informed readers how to tell the European carp from crucian or golden carp and described the 'geographical races' of the species. The evidence for prior introductions came mostly via a letter from G.P. Whitley (Curator of Ichthyology at the Australian Museum), and can be summarised as: David Stead introduced them to the Prospect Reservoir in 1907 and 1908, some fish were later released into Lake Windermere [sic] and

Centennial Park in Sydney, and some were kept in an aquarium in Taronga Park Zoo. The Committee rejected the notion that common carp had been widely introduced into Victoria and Tasmania between 1860 and 1890 'in quantity' (Wiltshire *et al.* 1962a, p. 11), and held that it was more likely those fish were crucian and golden carp. In addition to the few locations around Sydney, the Committee acknowledged the presence of common carp in 'a small number of irrigation ditches or channels in the Riverina' (Wiltshire *et al.* 1962a, p. 11) – presumably referring to the Yanco strain of carp – and the ones in the Melbourne Botanic Garden ponds.

From evidence presented, the Committee concluded that the species would thrive in most parts of Victoria except the alpine areas, because of the cold. Furthermore, although in Europe carp seemed to be restricted in the types of water bodies in which they can live and grow, in the US they had adapted to most types of water, flowing or still. Not only that, their tolerance of low dissolved oxygen levels, pollution and other water quality extremes meant that they were able to exist under a wide range of conditions. They were hard to catch when angling, and so often came to dominate the fish fauna of the US lakes to which they had been introduced.

The Committee was told of the impossibility of confining carp to a pond, reservoir or lake. They moved around naturally during floods and it was theoretically possible that they could disperse as eggs stuck to the legs or feathers of water birds. Deliberate human-mediated dispersal of carp could happen through anglers using small carp as live bait – which was well known – or people just moving them around for the hell of it. The Committee had to look no further than the actions of Pratt himself. He gave evidence that he had transferred carp from Prospect Creek to Boolarra and had already sold £700 worth of carp to all sorts of people, most of whom he could not name because he hadn't kept records. Based on the previous year's advertisement in the *Weekly Times*, when carp were offered at £20 per hundred, 3,500 carp may have been sold. Later, Pratt admitted to selling around 10,000 carp, and where they ended up was anybody's guess (Wiltshire *et al.* 1962b).

Pratt and especially Scott made contradictory arguments, but consistently used the logic that if carp was going to become a pest, it would have done so by now. One argument was that since carp had been in Prospect Reservoir for so long without having spread widely into streams it was not likely to spread in the future, especially if confined to ponds, dams or suchlike. The other argument, as we have said, was that carp *had* been widely introduced into flowing waters in Australia but had not come to dominate, and so future introductions would produce the same result. The Committee was unconvinced that either argument was evidence of a lack of a future takeover by carp.

A Czech fish and carp expert, Dr Vojtech Hollý, gave evidence on several aspects of carp biology, including diet. Butcher and Pratt did not fully agree with him or each other on what carp ate, but there was general consensus that carp could disturb aquatic vegetation through their feeding and this may, in some cases, lead to the death of those plants and to increased turbidity. All agreed that under the right conditions carp could grow quickly, and faster in Australia (2–3 lb [900–1300 g] per year) than Europe (1–1½ lb [450–650 g] per year). They could also reach large sizes in time: a 70 lb [31 kg] German carp was mentioned. Butcher claimed that large females (e.g. 15 lb [6.8 kg]) produced upwards of 2 million eggs, whereas Pratt put the number of eggs for a similar-sized fish at around 750,000 eggs. But there was consensus that a very large proportion of young was lost to predation, which was common for all fishes. The Committee came to the conclusion that, despite losses of young to predation, given its hardy nature European carp would thrive under the conditions in most Victorian water bodies. Evidence was stacking up against allowing the introduction of European carp into Victoria. But more was to come.

There was much disagreement on what impact European carp had on water bodies and the other fish that inhabited them. Arguing for the negative, Butcher's case against carp relied heavily on the US experience. US experts submitted that the usual pattern following carp introduction to a water body was destruction of aquatic plants, increased turbidity and the loss of native fishes. The elimination of carp apparently reversed the pattern. Arguing for the affirmative, Pratt claimed that other species of fish coexisted with carp because of dietary differences and that carp's effects on water quality were exaggerated. The Prospect Reservoir was a case in point: despite carp being in the water body for at least 50 years, aquatic plants were present, carp coexisted with other fish species and the water was clear. Hollý and L.G. Roth, garage proprietor and member of several regional development committees, gave further evidence for the affirmative case, saying that the environmental effects of carp were overstated and were similar to the effects of many other fish species when they occurred in large numbers.

The Committee was told in no uncertain terms, by those on the negative side, that once carp had got into Victorian waterways it was very unlikely they would ever be able to be totally eradicated. The US experience had been that carp eradication from large lakes and dams was nigh impossible, and eradication had only been achieved in small water bodies. Furthermore, the process of eradication had to be prosecuted intensively using multiple methods, and the associated labour and monetary costs were enormous.

The Committee then heard that the demand for European carp was limited. Newly arrived European immigrants and the Jewish community were the main markets, but demand by the former waned with time, especially among the children of immigrants. The Australian public, as a rule, was not adventurous in its fish-eating habits and was unlikely to take to carp enthusiastically. Therefore, there was little justification for risking environmental devastation by introducing a species for which there was little demand.

The Committee was acutely aware that evidence for the effects of carp in Victoria waters specifically was non-existent and therefore it had to rely on evidence from Europe and the US, which was frustratingly – but not surprisingly – contradictory. Evidence from the former suggested that carp might not be a problem, whereas evidence from the latter was decidedly the opposite. The Committee ultimately came down on the side of caution, based on evidence of the more recent and largely negative experiences in the US. It also raised the spectre of other disastrous introductions to Australia from Europe, like the rabbit and European or redfin perch. Butcher's recommendation was that the European carp be declared a 'noxious fish': possession and introduction of the species would be illegal and all individuals of the species, at any life stage, should be seized and destroyed. The new legislation, he argued, should cover natural and private water bodies, to avoid the loophole that had let Pratt get away with selling carp to individuals to stock their farm dams.

Pratt was, of course, far from happy. He attempted to get compensation for Boolarra Fish Farms in the event the noxious fish legislation went through. After considering the situation, the Committee demurred (Wiltshire *et al.* 1962a).

In May 1962 emergency legislation was passed – the *Noxious Fish Act* – which required anybody possessing European carp to notify the Department in writing. The Department was given the power to destroy carp wherever they were found and to impose a $1,000 fine on anyone who failed to comply.

Round 1 was won by Alf Butcher and the Victorian Department of Fisheries and Wildlife. But the win was a pyrrhic victory. In many ways, the hard work was only just beginning.

CARP-KILL

Following the government inquiry, Butcher wrote a short booklet – *Why Destroy the European Carp?* – in which he justified the State Development Committee's recommendation that carp be declared a noxious species. The booklet also clearly stated what he had learnt from his US trip and the potential for carp to be a disaster if it got into and spread in Victorian waterways (Butcher 1962). In 1960 and even early in 1961, Butcher and others at the Commonwealth–State Fisheries Conference were prepared to consider contained, limited experiments on carp in Victoria to assess its effects on native animals and plants. But after his trip to the US, Butcher did a complete turnaround. In his view, there was no longer any justification whatsoever for the introduction of European carp to Victoria. The Department advertised the new legislation in many regional and metropolitan newspapers:

> *It is an offence against the* Fisheries (Noxious Fish) Act 1962 *(No. 6885)*
> *for any person to keep or to release into any water in Victoria any European*
> *carp (*Cyprinus carpio *Linnaeus) or the eggs of such fish. Penalty – Not less*
> *than £100 and not more than £500. Any person owning or possessing such*
> *European carp must forthwith give or cause to be given written information*
> *of the existence and location of these fish to the Inspector of Fisheries, Fisheries*
> *and Wildlife Branch, 605 Flinders Street Extension, Melbourne. Failure to*
> *give such information can incur a penalty of £500. No penalty will be imposed*
> *where –*
> *a) That information is given to the Inspector of Fisheries by 31/5/62;*
> *AND*
> *b) The European Carp or the eggs of those carp are not removed from the water*
> *in which they now are until the arrival of officers from the Fisheries and Wildlife*
> *Branch.*
> *A. DUNBAVIN BUTCHER, Director of Fisheries and Wildlife (*Herald, *5 May*
> *1962).*

In response to the declaration of carp as a noxious species and in the wake of Butcher's booklet, Pratt and the Boolarra Fish Farms had a last swipe at the Department. They advertised

"Please Help to Destroy"

THE TRUE

EUROPEAN CARP

2/6 lb.

AVAILABLE AT:

BOOLARRA FISH FARMS

BOOLARRA IN GIPPSLAND.

Fig. 5.4. Boolarra Fish Farms' advertisement following the *Noxious Fish Act.*
Source: Redrawn from *Australian Jewish Herald*, 6 April 1962, p. 18.

carp for sale in the *Australian Jewish Herald* in Melbourne, riffing off Butcher's title *Why Destroy the European Carp?* (Fig. 5.4). They were not going to concede easily.

The carp-kill program ran from May 1962 to almost the end of the year. It was a mammoth task. There was no record of the people that Boolarra Fish Farms had sold carp to, so the distribution of carp in Victoria was unknown. The Department had to rely on farmers' declarations, but many could not remember how many fish they had bought from Boolarra Fish Farms (Pribble 1980). It turned out that carp had been sold all over Victoria and had been sold and shipped to several people in Stowport, Natone and Glance Creek, Tasmania (Inland Fisheries Commission 1975).

We have not managed to track down the records of all the dams poisoned around Victoria, and neither did Jim Pribble in 1980, but his Carp Program Annual Report for 1979/1980 has details of the carp-kill eradication in South Gippsland in an appendix (Pribble 1980). In that region alone, Boolarra carp were released into at least 716 dams. The dams essentially covered the whole region, from the coast to the Great Dividing Range. Small individual farms usually had only one dam stocked with carp, whereas in larger communities such as Leongatha (31 dams), Mirboo North (49), Glen Alvie (30), Bena (31) and Loch (22) often many dams had been stocked. Eighteen dams at Boolarra were stocked with carp.

At least in South Gippsland, 15 different chemicals were considered as candidates for poisoning. Assessments were based on safety, the amount needed, the cost and whether there was residual toxicity. In the end, sodium pentachlorophenate (sold as Santobrite) and lime (Limil) were mostly used. Santobrite directly killed the fish, whereas lime took the pH of the water above lethal levels.

In his book *Heads and Tales*, Victorian fisheries officer Jack Rhodes described the process of poisoning a dam. He said it was easy to tell which dams had carp in them, because 'if there was no one at a farm when we arrived to treat a dam or dams, we could, as did other teams, unerringly select the right dams because of the turbidity of the water' (Rhodes 1999, pp. 177–178). Rhodes implied that Boolarra Fish Farms actively 'hawked' carp around Victoria, and that if a farmer did not buy fish, some were nevertheless thrown into a dam if it ran alongside the property's access road. This statement was based on observed changes in turbidity in the dam following the hawker's visit, and the fact that when those particular dams were poisoned, carp were always present.

Poisoners worked in teams of two or three, based at a number of centres around Victoria. Each team pulled a trailer full of lime (or another chemical) and a boat with an outboard motor, at times across rough terrain which sometimes made the task dodgy, to say the least (Rhodes 1999). They estimated the volume of water in each dam and measured the pH of the water, then calculated the amount of chemical needed for an effective kill. The chemical was poured into the turbulent water created by the boat's propeller, which was faced towards the dam centre. With lime, the water's pH needed to be raised to 11 for a minimum of 15 minutes, after which fish started coming to the surface. Of course, the chemicals killed all fish in the dams, including a variety of native fishes and trout, redfin perch and goldfish (Pribble 1980). The farmers were given rainbow trout in exchange for the carp. It was estimated that more than 15,500 carp had been released into dams in South Gippsland alone. The campaign in this region killed more than 8,500 carp: 1,100 extracted from the dams before the poisoning began and 7,480 directly poisoned. Some carp had already been eaten.

By the time the Victoria-wide carp-kill program wound up, it had involved 37 departmental employees, taken 2,000 people-days of effort and cost a whopping £20,500 (about $50,000) at the time (Wharton 1971). Accounting for inflation, this equates to about $700,000 in today's money. In all, more than 1,300 dams were poisoned. Even the carp in the Melbourne Botanic Gardens shared the fate of their finny compatriots in farm dams. Approximately double the number of South Gippsland dams were included in the Victoria-wide carp-kill program. Based on this number of dams (and no doubt the many unreported dams), there must have been a lot more than the 3,500 carp Pratt claimed had been sold to farmers. It would have been more like 30,000–40,000. This suggests that either Pratt's memory was very poor or that he had been economical with the truth in his evidence to the State Development Inquiry. For the carp-kill program as a whole, 200 dams were re-tested at a later date, and 'to our great pleasure, not a single live carp was found in any of the dams visited' (Wharton 1971, p. 7).

*****

It would seem that the poisoning had been a complete success.

6

Biology

Mating groups of one female and several males (but several females per one male were also observed) circled ferociously, often disturbing the surface or with dorsal parts out of the water (Balon 1995a, p. 18).

KNOWLEDGE OF CARP BIOLOGY IS FUNDAMENTAL IF WE HOPE TO MAKE sense of how carp has proliferated in Australia – and for that matter, all around the world – how they have affected our rivers, lakes and wetlands and how people have attempted to control and manage them in the last 60 years or so. If a silver bullet is found or if we can locate the carp's 'Achilles' fin' – and thus remove the scourge of carp from Australian waters – it will be the carp's biology that provides the clue to either or both. In our exploration, we must keep in mind aspects of the freshwater environments in Australia that are unique to this country and those that are shared with other continents. There is no doubt much we can learn from the study of carp elsewhere; dismissing what others have found would be foolhardy and a waste of time and money. But we would also be wise to take nothing for granted, and remain vigilant in case the Australian situation throws up unusual insights and novel opportunities.

In this chapter, we give general biological principles about the common carp, based on what is known globally, then we narrow the spotlight onto what we know from Australia. It is fortunate that since the Boolarra carp escaped 60 years ago, in what we have called the Boolarra Event Horizon or BEH, there has been much research and substantial gains in our understanding of the species' biology in general and its ecology in particular. This has also occurred on continents that see the common carp as an invasive species where the priority is control and management, and even eradication, like North America and New Zealand (Vilizzi *et al.* 2015).

To whet the appetite, we begin this chapter with observations of the maximum age and size common carp can attain, globally and in Australia. Then we move on to the life cycle of the species, at what age it matures, when, where and how it spawns, and aspects of its early life. We next focus on habitat, movement and environmental tolerances, before completing our exposé of carp biology with food and feeding.

HOW OLD AND BIG CAN CARP GET?

It is always with a healthy dollop of scepticism that one should approach claims of maximum ages and sizes of fish. Fortunately, much of our work has been done for us via several reputable sources: FishBase, a compendium of all things fishy (Froese and Pauly 2023); a review in the peer-reviewed *Journal of Fish Biology* of 284 studies, spanning 91 years and 50 countries by Lorenzo Vilizzi and Gordon Copp (Vilizzi and Copp 2017); and other studies and reviews of the biology and population dynamics of carp closer to home (Hume *et al.* 1983; Brumley 1991, 1996; Vilizzi and Walker 1999a; Koehn *et al.* 2000; Brown *et al.* 2003; Sivakumaran *et al.* 2003; Smith and Walker 2003, 2004a, b; Koehn 2004; Vilizzi 2018). The ages of fish are usually determined using annual growth rings (like trees) that are deposited in hard parts such as scales, bones, fin rays, vertebrae and, most commonly used, ear bones,

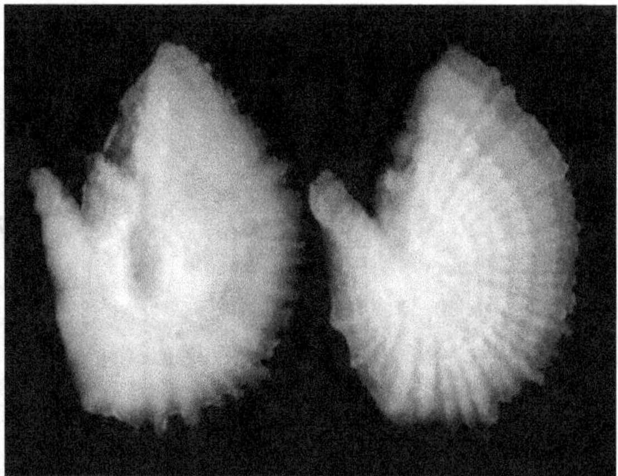

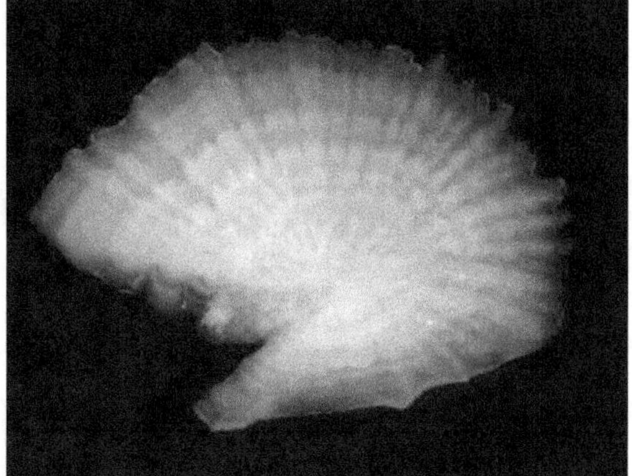

Fig. 6.1. Otolith structures of common carp used in ageing. Source: An Vi Vu.

otherwise known as otoliths (Fig. 6.1) (Vilizzi 2018). In the past, scales were used to age common carp; however, they have proved unreliable because they can be damaged or even lost and replaced.

There have been several anecdotal reports of carp that lived for more than 40 years, and even one that lived in captivity for more than 140 years (Vilizzi and Copp 2017). The UK's 'Two-Tone' supposedly was 45 years old when he died, and had been caught many times over the years (Cherrell 2022). Equally spectacular and equally anecdotal, common carp reaching 32 kg in Italy, 37 kg in North America and 57 kg from the Dnieper Delta at the Black Sea in Europe have been claimed.

More reliably, FishBase gives the maximum length, weight and age of common carp as 120 cm, 40.1 kg and 38 years, respectively. Vilizzi and Copp (2017), in their extensive review, cite more modest extremes of just over 100 cm in length and 27 kg in weight. They also found that no individual carp older than 19 years had both its age estimated and length measured. Estimated maximum reported ages of 30+ years (without lengths) for some populations of carp do exist and come from credible sources (Bajer and Sorensen 2010; Bajer *et al.* 2015; Weber *et al.* 2015). It seems that carp can live for several decades under the right circumstances.

Although there is substantial sex size variation for any age, females on average grow larger than males (Vilizzi and Walker 1999a; Vilizzi and Copp 2017). This sexual dimorphism is common globally and in Australian carp (Brown *et al.* 2005). Female fish are often larger than males because the number of eggs typically correlates with size (Fig. 6.2). Males, on the other hand, can produce enormous quantities of sperm even when relatively small. Furthermore, some research suggests that females

Fig. 6.2. A fat female carp. Source: Clayton Sharpe.

are fatter for any given length (Brown *et al.* 2003), but this is by no means always the case (Vilizzi and Walker 1999a).

The maximum age that we know of for a common carp in Australia was a male from Gippsland in south-eastern Australia, at 32 years (Brown *et al.* 2003). Carp older than 25 years from the Murray River at Barmah, Kerang Lakes, Barwon River and Gippsland Lakes were sufficiently common to suggest that, under the right conditions, fish can live as long under Australian conditions as they can elsewhere in the world (Brown *et al.* 2003). One of the most comprehensive studies on carp in Australia, by Lorenzo Vilizzi and Keith Walker (1999a) in the lower Murray River, found the largest carp of the 603 fish collected was a 7-year-old female, weighing almost 7 kg. Their oldest fish was a 15-year-old female that weighed just 5.5 kg. Overall, Australian carp older than 10 years are probably outliers, with most adult carp being 3–11 years old and 300–600 mm caudal fork length long (Brown *et al.* 2003).

LIFE CYCLE

For all the reasons we have seen in previous chapters, the life cycle of the common carp is one of the best studied of all freshwater fish, apart perhaps from the salmons and trouts. However, whenever we wish to generalise about, for example, spawning behaviour, spawning-related movements and spawning habitat, the environment in which the fish lives must be considered.

Global overview of the life cycle of the common carp

The carp normally matures at 2–5 years of age and spawns during the warmer months of the year. In the UK and Europe, this is June–July when the water temperature is 15–20°C (Maitland and Linsell 2006). There have been reports of carp maturing as early as 3 months (in South-East Asia) and as late as 5 years (in Russia) (see Hume and Pribble 1980). Some populations of carp spawn for many months each year and at temperatures as high as 28°C (Panek 1987). Fish almost always lay eggs on aquatic plants or flooded terrestrial vegetation (Fig. 6.3), often in shallow, slow-flowing or still habitats, and, when in rivers, often associated with rises in flow.

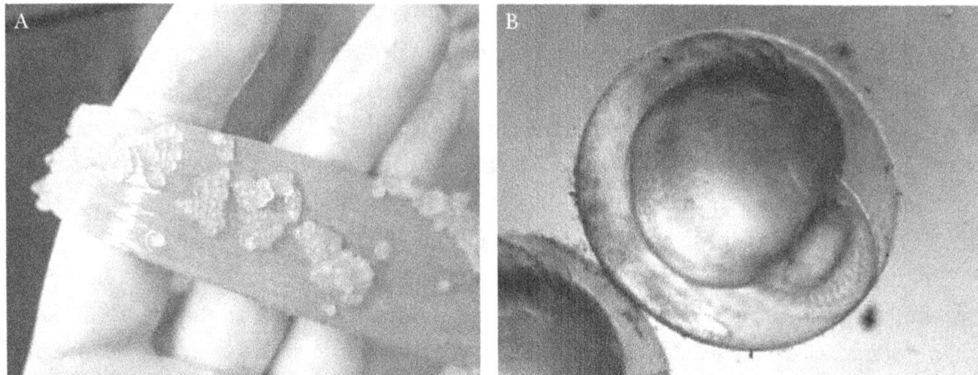

Fig. 6.3. (A) Carp eggs on vegetation. Source: Keith Bell. (B) Four-day-old carp egg, with signs of notochord development. Source: Katie Doyle.

Fig. 6.4. (A) Adult carp entering a shallow inundated floodplain. (B) Adult carp with their backs protruding in shallow water on inundated floodplains. (C) Spawning aggregation of carp in shallow water of the Lachlan River, New South Wales. Source: Warren Chad.

During spawning, multiple males may chase a female, repeatedly bumping her in an effort to get her to release her eggs (Scott and Crossman 1973). Carp spawning can be quite a noisy affair (Balon 1995a) with aggregations of excited spawning fish splashing around in the shallows, seemingly oblivious to all around them (Fig. 6.4).

As a rule, the common carp can be classified as a multiple batch, asynchronous spawner, which means that fish may spawn more than once in a season, depending on environmental conditions, but not all fish spawn all at once (Crivelli 1981; Smith and Walker 2004a; King *et al.* 2013). The number of eggs a female produces depends on her size: from approximately

50,000–300,000 per kg of body weight and 12,000–1,540,000 per fish (Linhart *et al.* 1995; Weber and Brown 2012). It is not uncommon for females 70+ cm long to produce well over a million eggs (Swee and McCrimmon 1966; Sivakumaran *et al.* 2003; Bajer and Sorensen 2010). The diameters of fertilised eggs are approximately 1.2–1.4 mm and are sticky so they stay attached to plants and avoid what can be poor water quality, and especially low levels of dissolved oxygen, on the substratum (Linhart *et al.* 1995). Parents do not physically care for their eggs or young fish post-hatch (Hume and Pribble 1980).

Because of their propensity to make use of inundated habitats, common carp often make spawning movements into such areas (Bajer and Sorensen 2010) (Fig. 6.4), and there is evidence that in some cases individual fish return to the same locations to spawn each year (Banet *et al.* 2022). Once spawning is complete, adult carp usually return to deeper floodplain habitats, if available, or enter lagoons, anabranches or river channels to avoid becoming stranded in shallow waters. Dispersal of young carp from floodplain spawning habitats into the main stem of rivers is commonplace (Crook and Gillanders 2006). Bajer and Sorenson (2010) hypothesise that common carp have evolved to take advantage of ephemeral habitats that may be less attractive for other species of fish, but give young carp a better chance of survival. They further suggest that this is consistent with the seasonably variable aquatic environment in the region in which they evolved.

The sporadic and variable nature of inundations means that conditions are not always favourable for spawning and/or recruitment (King *et al.* 2003), which has led to carp being classified as having a 'Periodic' life history (Winemiller and Rose 1992; Vila-Gispert and Moreno-Amich 2002): there are years in which recruitment is good and other years in which recruitment is poor (Humphries *et al.* 2008). Good recruitment is typically linked to warm temperatures, the presence of flooded vegetation, abundant food supply and a relative absence of predators, both in the wild and under aquaculture conditions (Swee and McCrimmon 1966; Crivelli 1981; Kanitskiy 1983; Bajer *et al.* 2015; Lechelt and Bajer 2016; Maiztegui *et al.* 2019).

Carp embryos hatch after 3–6 days, depending on temperature, and the optimal temperature range for survival, hatching and growth ranges from 17–26°C (Wieser 1991; Sapkale *et al.* 2011). Newly hatched young carp stay close to plants, again most likely to avoid what could be poor water quality on the substratum (Swee and McCrimmon 1966). After about 5 days, the young fish's swim bladder fills and the now free-swimming 7–10 mm carp has used up its yolk sac and begun to feed on aquatic prey (Fig. 6.5) (Balon 1995b). The young carp's

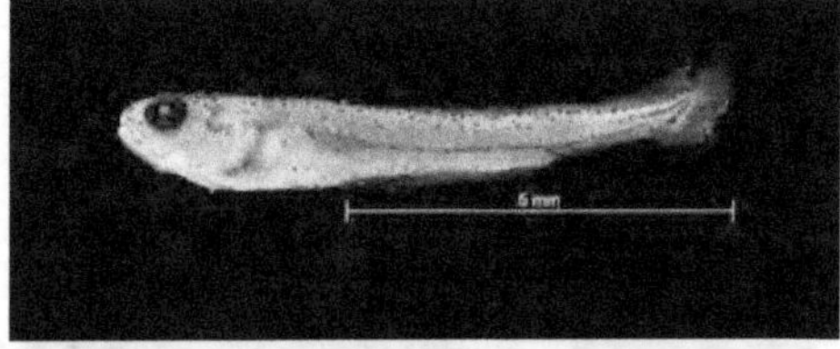
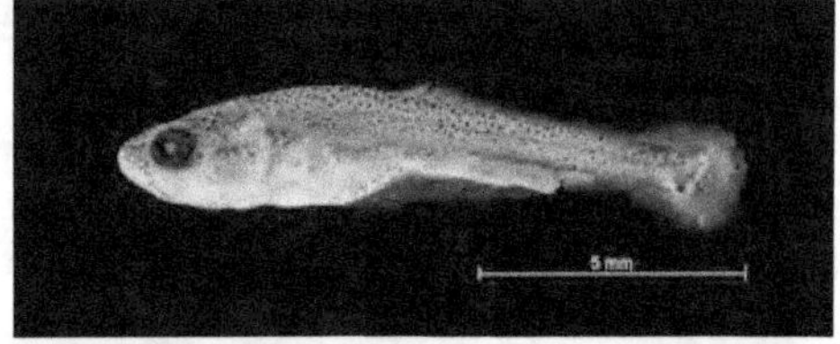
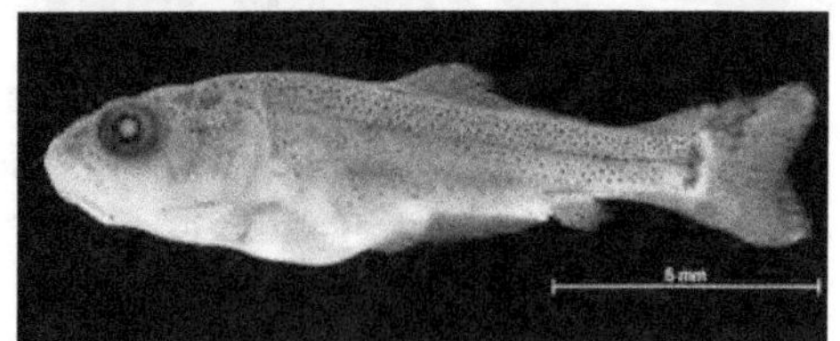

Fig. 6.5. Morphological development of carp from approximately 7 days post-hatch to approximately 25 days post-hatch, just prior to becoming a juvenile. Source: John Trethewie.

swimming ability improves at this time so that it is able to actively pursue prey (Tonkin *et al.* 2006) and its gills take over as the primary respiratory organ (Balon 1995b). Juvenile common carp are apparently very tolerant of high water temperatures, although this depends on the temperature at which they were acclimated. For example, Opuszynkski *et al.* (1989) found that juvenile carp could survive up to 43–46°C and optimum growth occurred at 38°C, although this is certainly not the case in all studies that have investigated thermal tolerances (see section **Habitat, movement and environmental tolerances**).

Carp become juveniles at 20–25 mm long, approximately 20–30 days after hatching (Vilizzi and Walker 1999b). While feeding exclusively on their yolk sac, the mouth is oriented downwards. It orients to be more horizontal by the time the fish – now a larva – begins feeding externally, before reverting to a downward oriention in juveniles (Hoda and Tsukahara 1971). This reflects the change from feeding internally to externally, and from feeding on planktonic zooplankton as a larva to feeding on bottom-dwelling prey as a juvenile (Adzhimuradov 1972; Balon 1995b; Vilizzi 1998). Once carp become juveniles, they pretty much behave and look like mini-adults.

REPRODUCTION OF CARP IN AUSTRALIA

Much of what follows relies on the empirical studies and reviews by Andrea Brumley, John Koehn, Paul Brown, Kathiravelu Sivakumaran, Ben Smith and their respective colleagues (Brumley 1996; Koehn *et al.* 2000; Brown *et al.* 2003, 2005; Sivakumaran *et al.* 2003; Smith and Walker 2004a, b). In the southern Murray–Darling Basin, males and females have been found to mature before they reach their first birthday (at 15 cm and 25 cm total length, respectively) (Hume *et al.* 1983), whereas in Gippsland, Victoria, males were mature at the end of their first year of life and females at a year older (Brumley 1996). Further west and north, in Campaspe irrigation canals and Barmah-Millewa Forest, although some fish of both sexes were mature before their first birthday (at 23–28 cm, caudal fork length), the percentage that were mature increased with age, with most mature by 4 years of age (31–39 cm caudal fork length) (Brown *et al.* 2003). From some regions of New South Wales, male carp mature at 2–3 years old (at 30 cm) and females at 3–4 years old (at 35 cm) (Brown 1996). In Tasmania's Lake Crescent, where temperatures are particularly cool and fish grew slowly, the earliest males matured at 3 and females at 4 years old, although some females took a whopping 7 years to mature (Diggle *et al.* 2012). Sex ratios of carp depend on location, with 1:1 in some places (Brown *et al.* 2003), whereas elsewhere females (Smith 2004) or males can dominate (Stuart and Jones 2006a). There are some indications that females may become rarer as populations age.

In Australia, carp too is mostly a multiple batch, asynchronous spawner, like its overseas cousins, with a highly variable breeding season, although breeding normally commences in spring when eggs mature (Sivakumaran *et al.* 2003; Smith and Walker 2004a). Carp larvae have been collected – indicating that fish have spawned – from several locations where water temperatures were as low as 13–16°C (Hume *et al.* 1983; Humphries *et al.* 2002; King *et al.* 2003; Smith and Walker 2004a, b). Some carp in Lake Crescent, Tasmania, were found spawning at a chilly 11°C; however, no eggs survived to recruit at this temperature (Inland Fisheries Service

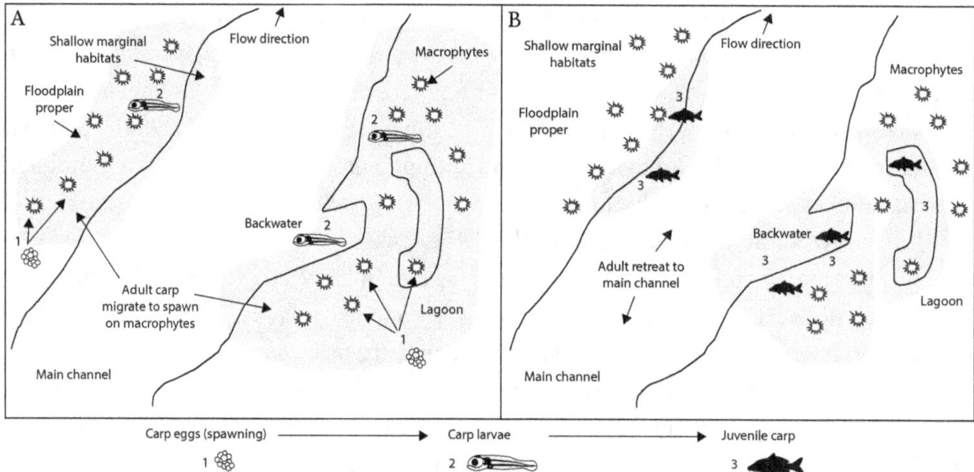

Fig. 6.6. Schematic of a river channel and associated floodplain to indicate movements of common carp (A) into and (B) out of habitats before and after breeding. Source: Katie Doyle.

2004). Observations from the Murrumbidgee Irrigation Area indicate that carp spawned when water temperatures were between 19 °C and 29 °C (Adamek 1998) and in Victoria when water temperatures were between 17 °C and 25 °C (Hume *et al.* 1983). Normally, it seems that numbers of hours of daylight needs to exceed 10 before spawning can begin (Brown *et al.* 2005).

In South Australia and at Barmah-Millewa Forest, water temperatures remain warm and carp have been known to spawn across 6–7 months of the year (Sivakumaran *et al.* 2003; Smith and Walker 2004a, b). Females may, therefore, spawn 2–3 batches of eggs during that period. In the Tasmanian lakes Crescent and Sorell females have been known to produce 2 batches in a season (Donkers 2003a). So, depending on the location and the conditions, common carp in Australia may spawn from early spring to autumn (September–April), especially if the temperatures are warm and flows increase to cover normally dry vegetation. Carp have been found spawning during droughts and floods (Humphries *et al.* 2002; King *et al.* 2003).

Carp in Australia usually spawn during the daytime (Hume *et al.* 1983), although this is at odds with what has been reported in Europe and Asia, where spawning during the night or early morning is common (Sarig 1966). As they do overseas, Australian adult carp tend to form large – often noisy – schools that swim into floodplain habitats and backwaters and edge habitats of the main channels of rivers, or in the shallow regions of lakes (King *et al.* 2003; Crook and Gillanders 2006; Stuart and Jones 2006a) (Figs 6.4 and 6.6).

During the Millennium Drought in south-eastern Australia, suitable spawning habitat was scarce in the main channel of most rivers, and so floodplain forests such as those at Barmah-Millewa were apparently important locations for carp spawning and recruitment (Stuart and Jones 2006a). Observations have led some scientists to suggest that there are hot-spots for carp spawning and recruitment, and this knowledge might be exploited in the search for ways to control carp (Brown *et al.* 2005; Macdonald *et al.* 2010). The increased availability of suitable habitats for spawning and recruitment during floods means that recruitment success is often greater in flood years (Humphries *et al.* 2008).

As with common carp elsewhere in the world, the total number of eggs produced annually depends on the female's size, age and condition (Hume *et al.* 1983; Sivakumaran *et al.* 2003). In the Victorian Carp Program study, total numbers of eggs per female ranged from 45,000 for a 39.5 cm, 1.22 kg fish to 1,500,000 for a 62 cm, 6.4 kg fish (Hume *et al.* 1981). Another study found that total egg numbers ranged from 12,000 eggs for a 39 cm fish to 1,540,000 eggs for a 78 cm fish (Sivakumaran *et al.* 2003). In the latter study, the average total number of eggs per female was approximately 750,000 and the average number of eggs per kg of fish was 162,000 (range 75,000–262,000 per kg). The number of eggs per kg of fish was not related to size or age. Egg diameter ranged from 0.4 mm to 1.6 mm and was weakly positively related to fish length and weight (Sivakumaran *et al.* 2003).

EARLY LIFE OF CARP IN AUSTRALIA

If spawning is associated with flooding, then young carp must move out of flooded areas before they dry, and go into more reliable bodies of water: the main channel and anabranches of rivers, slackwaters and off-channel lagoons (billabongs or oxbow lakes) and inundated, relatively permanent floodplain habitats. Often young carp do this *en masse*, which provides a degree of protection from predators, although how this is coordinated is not understood (Vilizzi and Walker 1999b). Ideally, they move into environments that have aquatic plants (because juveniles too like plants), are shallow, warm, full of food and with a minimal risk of drying (King *et al.* 2003; Brown *et al.* 2005; Stuart and Jones 2006a). Juvenile carp tend to prefer off-channel environments if they are available (Gehrke and Harris 1996), but we have collected them in the main channel of the Campaspe and Broken rivers over many years (Humphries *et al.* 2002, 2008). These rivers are not known for their extensive floodplains, but they do have large areas of slackwater in-channel.

Following flooding, growth in young carp is greatest as temperatures warm, food becomes plentiful and the density of fish is not too great (Hume *et al.* 1983; Vilizzi and Walker 1999b). Growth may be limited if food becomes scarce, but will recommence when food production increases again.

HABITAT, MOVEMENT AND ENVIRONMENTAL TOLERANCES
Habitat

Carp in Australia is found in a diversity of habitats: the main channels of rivers, associated floodplains, billabongs, swamps, irrigation canals, and natural and artificial lakes and reservoirs (Koehn *et al.* 2000; Stuart *et al.* 2006) (Fig. 6.7). It is rarely found in cool, rapidly flowing waters associated with montane sections of rivers (Faragher and Lintermans 1997). Although it occurs in unregulated and regulated systems, it makes up the greater proportion of fish weight in regulated river sections, especially the lowland reaches (Gehrke 1997; Gehrke and Harris 2000). Carp also seems to benefit and be abundant where humans have degraded natural systems (Driver *et al.* 1997; Gehrke *et al.* 1999). In Australia, carp can normally be found in freshwaters as

Fig. 6.7. Range of habitats in which carp are found in Australia. Source: Katie Doyle.

far downstream as the tidal limit. It will enter estuaries (Koehn *et al.* 2000) but can be susceptible to sudden changes in salinity (Buckmaster 1975), since its upper limit to salt concentration is normally 12,000–15,000 mg/L (Wieser 1991). There are occasional reports of carp in the ocean, but it is likely that in fact the fish were entrained in river plumes (Murdy *et al.* 1998).

Carp can be found at a range of depths, often associated with woody debris (but higher in the water column than Murray cod), closer to the bank than similar-sized native species (Murray cod, trout cod [*Maccullochella macquariensis*] and golden perch) and at relatively slow current speeds (Koehn and Nicol 2014). Due to its feeding mode, carp is usually associated with fine sediments such as sand, silt or detritus, which are conducive to the growth of aquatic plants. So, the fish tends to avoid areas with large bed elements like cobbles, boulders and bedrock.

Movement

The movement patterns of carp have been extensively studied in Australia and overseas, and some of the most comprehensive research has been local (Reynolds 1983; Mallen-Cooper *et al.* 1995; Koehn and Nicol 1998, 2016; Stuart and Jones 2006b; Jones and Stuart 2007, 2009; Conallin *et al.* 2016). The movement of carp was an important aim of one of the first studies of the species in Australia. Following the spread of carp into South Australia, Fred Reynolds was appointed by the South Australian government to study the biology of freshwater fish, including 'European carp' in 1974. Reynolds tagged fish in the lower Murray River, using Floy anchor tags from 1974–1978 (Fig. 6.8) (Reynolds 1976, 1983). Fish were recaptured by Reynolds'

Fig. 6.8. Floy tag in the dorsal region of a carp. Source: Katie Doyle.

team and by commercial and recreational fishers. A total of 5,268 common carp was tagged and 76 recaptured in the Murray River. A further 3,365 carp were tagged in Gurra Lakes (just downstream of Renmark, South Australia); 252 were recaptured in the lakes and 17 in the river. Recaptures of river-caught fish amounted to 1.4% tagged and recaptures of lake-caught fish 8%, yet Reynolds drew conclusions on carp movement from these recaptures. With this number of recaptures, it is possible to say how much the recaptured fish may have moved, but impossible to know how much the non-recaptured 98.6% of the river fish and 92% of the Gurra Lakes fish moved. That aside, Reynolds' work has provided insight into the distances that carp and some native fish, especially golden perch, are capable of moving.

Most carp that Reynolds caught were <200 mm long (0+ or 1+ fish). Some of his river-tagged fish moved into the lakes, and *vice versa*. And some lake fish transferred to the river returned to where they had been captured. He found that the maximum upstream and downstream distance recaptured carp had moved was 80 km and 73 km, respectively, and that few fish moved more than 10 km (Reynolds 1983). He concluded that carp did not migrate, *per se*. Time of year and flow were apparently unrelated to movement distance or direction. His results of the supposed lack of carp movement – based on his recaptures – was reported extensively at the time (e.g. Australian Fisheries Newsletter 1976a; Reynolds 1976). His conclusions, however, were seemingly at odds with the rapid spread of carp that we know occurred after the BEH and presumably was happening during his tagging study.

More recent studies monitoring movement through fishways, using radio transmitters embedded in fish, indicate a more nuanced story. In the early 1990s, Martin Mallen-Cooper and colleagues recorded fish passing through the Torrumbarry fishway in the mid-Murray region (Mallen-Cooper *et al.* 1995). From February 1991–July 1992 they recorded 16,000 carp, with movement apparently related to rising water temperatures rather than to increase in flow. Indeed, the greatest numbers of fish movements occurred during the lowest flows, once temperatures reached 20°C. John Koehn and Simon Nicol also showed the propensity for carp to move and at similar temperatures (Koehn and Nicol 1998, 2016). Carp moved most commonly during the morning and to a lesser degree in the afternoon, but rarely at dawn or dusk or during the night.

These studies are supported by more recent work of Matt Jones and Ivor Stuart and of Rex Conallin. All showed that carp is able to move considerable distances upstream and downstream and from the main channel of rivers into floodplain environments, especially but not exclusively during the warmer months of the year (Mallen-Cooper *et al.* 1995; Koehn and Nicol 1998, 2016; Stuart and Jones 2006b; Jones and Stuart 2007, 2009; Conallin *et al.* 2012). Movement distances are highly variable and many carp do not move much at all. For instance, the median distance moved by 47 radio-tracked carp was approximately 2 km, although the odd carp did move more than 100 km (Koehn and Nicol 2016), echoing earlier results (Koehn and Nicol 1998).

A later study of 293 recaptures from more than 3,000 tagged carp in the Barmah-Millewa Forest adjacent to the Murray River between November 1999 and July 2001 showed that 66% of fish were recaptured within 1 km and 85% recaptured within 29 km of where they were tagged (Stuart and Jones 2006b). However, a few fish travelled long distances: one fish moved 203 km upstream in 423 days and another an astonishing 890 km downstream in 1,107 days. Male carp apparently move on average twice as far as female carp, with large females moving the least. River flow or water temperature apparently do not influence the distance carp move (Jones and Stuart 2009). The Tasmanian experience in lakes Crescent and Sorell showed similar variation in movement distance among individuals (Diggle *et al.* 2012).

In summary, studies show that the majority of carp stay close to home, but a small number may move large distances and do so quite quickly. In reality, Reynolds' results and those from more recent studies are not dissimilar. It is probable that Reynolds missed recapturing the 'travellers' in his populations.

The propensity for many carp to stay in one place, or in some cases to return to a place once they have moved, is a type of site fidelity (Crook 2004; Stuart and Jones 2006b; Jones and Stuart 2009; Piczak *et al.* 2023) or perhaps territoriality (Diggle *et al.* 2012). Dave Crook radio-tracked 15 carp in the Broken River, south-eastern Australia, some of which were released where they were caught and some translocated (Crook 2004). Of those carp that were not translocated, 4 of 5 maintained home ranges near where they were caught and released. Two of the 4 carp translocated upstream and 3 of the 5 carp translocated downstream moved back to the spot where they were originally caught. The fish that did not return, established home ranges in other locations. Radio-tracked carp in Barmah-Millewa Forest and the adjacent Murray River showed similar levels of site fidelity, with some notable exceptions (Jones and Stuart 2009).

Prior to the onset of high river flows in spring, common carp tend to congregate near access points to off-stream spawning habitat, and are often the first large species to move into these areas (Stuart and Jones 2002). They are often the last to leave, and many become trapped and die. Radio-tagged carp were found to move from the main Murray River channel into the adjacent Barmah floodplain and Moira lakes in response to moderate rises in flow, then further rises cued them to move into inundated floodplain habitats (Jones and Stuart 2009). Decreases in flow initiated movements back into the main river channel. Whether this can be classified as a spawning migration *per se*, since the movement can be local, but lateral from the main channel out into adjacent flooded habitat, is probably a moot point. But what is certain is that there is intentional and active movement to locate the best spots for spawning, which can be a useful way of catching fish in the act, so to speak (Stuart *et al.* 2006; Conallin *et al.* 2016).

Environmental tolerances

The carp's tolerance of poor water quality is one of the main reasons for it being one of the most globally dispersed freshwater fish species. There is little need to distinguish here between the tolerances of carp in Australia and the rest of the world, because the results are essentially the same. Interestingly, different strains and even different forms – fully scaled, mirror, linear or leather – have slightly different morphological traits, growth rates and environmental tolerances (Kirpitchnikov 1999).

Overall, carp has a broad tolerance to temperature, dissolved oxygen, salinity, pH and heavy metals (Table 6.1) (Crivelli 1981; Edwards and Twomey 1982; Opuszyński *et al.* 1989; Driver *et al.* 2005b). The upper and lower environmental limits that the carp tolerates depends on the region and acclimation (the conditions under which individuals were kept prior to experiments). They also depend on the life stage: adult carp are the most tolerant to environmental variables and early life stages are the least tolerant. Importantly, results vary from study to study, sometimes

Table 6.1. Lower and upper limits, preferences (where relevant) and sources of information for environmental tolerances of common carp.

Environmental factor	Lower limit	Upper limit	Preference	Source
Temperature (°C)	2 (Tas.)	41–46 (Europe, North America)	20–24	(Opuszyński *et al.* 1989; Eaton and Scheller 1996; Golovanov and Smirnov 2007; Diggle *et al.* 2012; Oyugi *et al.* 2012; Souza *et al.* 2022)
Dissolved oxygen (mg/L)	<1 (at 5°C); 0 for some hours at low temperatures; cannot gulp air	?	?	(Ott *et al.* 1980; Van den Thillart and van Waarde 1985; Stecyk and Farrell 2002; McNeil and Closs 2007)
Salinity (mg/L)	0	~15,000; can survive direct placement in 12,500; all died at 18,000	?	(Geddes 1979; Wieser 1991; Whiterod and Walker 2006)
pH	5	11	7–8	(Oyen *et al.* 1991; Diggle *et al.* 2012; Hellawell 2012; Heydarnejad 2012; Nasir et al. 2019)

? = unknown.

because of the fish subjects and their origins but often because few studies of environmental tolerances are carried out in a consistent way.

Scientists have studied the upper temperature limits of the carp much more than the lower limits (Table 6.1) (Wieser 1991). Overall, the common carp thrives in warm conditions and does not grow as fast or feed as much in cooler conditions (Maitland and Linsell 2006), but it nevertheless is able to cope with a broad range of temperatures. Oxygen dissolved in water is fundamental for the survival, growth and reproduction of the vast majority of aquatic animals. In the absence of dissolved oxygen, most animals will die – this is one of the main causes of fish kills in Australia in recent years (Australian Academy of Science 2019). Levels above 5 mg/L are usually desirable for freshwater fishes (Australian and New Zealand Environment and Conservation Council 2000). The carp is able to tolerate low levels, and even anoxia (0 mg/L), for a few hours. At low levels it will come up to the surface, not to gulp but to obtain enough oxygen from the oxygen-rich layer at the surface (McNeil and Closs 2007). It is also able to slow its metabolism and so consume less oxygen (Zhou *et al.* 2000).

Aquatic animals have to osmoregulate or control the balance of water and dissolved salts in bodily fluids. In Australian freshwater systems, salinity values <1,000 mg/L are considered healthy and values above that may have adverse effects on fish and other freshwater organisms (Australian and New Zealand Environment and Conservation Council 2000). The carp is able to tolerate salinity levels up to about 15,000 mg/L but will struggle after that. Reports of carp living in pure seawater are unlikely (Wieser 1991). pH is a measure of the acidity or alkalinity of water. Neutral water has a pH of 7, with values <7 being more acidic and values >7 being more alkaline. Most fish survive well in pH 6.5–9.0 but may experience poor health outside this range (Australian and New Zealand Environment and Conservation Council 2000). The carp can survive within the pH range 5–11 but it prefers 7–8, which means that it can survive quite happily in most uncontaminated Australian freshwaters.

FOOD AND FEEDING

Much of the impact of carp in Australia is related to its feeding and its role in freshwater food webs (Chapter 8). Here, we describe what is known about the feeding and diet of the species overall, then what we know about them in Australia. The summary draws mostly on studies by Crivelli (1981), Sibbing (1982; Sibbing *et al.* 1986; 1988), Michell and Oberdorff (1995), García-Berthou (2001) and Farag (Farag *et al.* 2014). Note: many studies of carp feeding and food have been in aquaculture settings or rice fields.

Feeding structures and food of the common carp

The adult carp does not have teeth in its mouth. Instead it has molar-like pharyngeal teeth, located further back in the area of the gills (Fig. 6.9). The pharyngeal teeth are located on the fifth gill arch, and include a chewing pad which allows crushing and grinding of food. Additionally, gill rakers (bony, finger-like structures on the inner edge of the gill arch) sieve out larger items that the carp may suck up from the water and sediment. This rejected

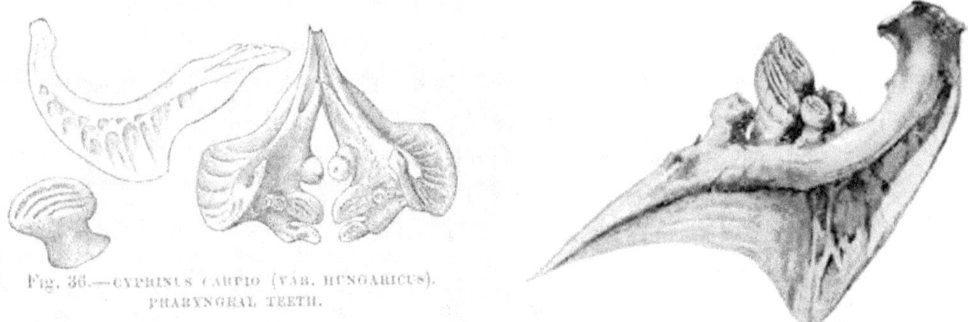

Fig. 6.9. Pharyngeal teeth of common carp. Source: Wikimedia Commons, CC BY-SA 4.0.

Fig. 6.10. Protrusion behaviours in common carp. (A) Resting state. (B) Partial mouth protrusion. (C) Closed mouth protrusion. Source: Katie Doyle.

material is spat out via the operculum. The material that gets through the rakers and ground up by the pharyngeal teeth is swallowed.

Carp is primarily an omnivorous bottom-feeder, but supplements its food from the water column when bottom-dwelling animals and plants are scarce. Fish will feed on soft aquatic plants (macrophytes), mostly dead ones, when animal prey is scarce, but this is not preferred, and fish will not grow if only eating plant material. Bottom-dwelling animals that are often in carp diets are zooplankton, chironomid or midge larvae (probably the dominant component of the diet), snails and small bivalves, mayflies and

Fig. 6.11. The upper lip (1) and corner of mouth (2) barbels on the right side of the mouth of the common carp. Source: Katie Doyle.

caddisflies, worms, grains (when available) and detritus (dead animal and plant material).

Many of the animals that the carp prefers, especially chironomids and worms, are associated with organic debris and detritus and so these last two may only be incidentally ingested. Feeding on the bottom normally involves sucking up material fairly indiscriminately, and sorting the good from the bad later. Suction is created through the fish's protrusible jaws: its premaxilla can be substantially extended to make a straw-like structure which creates negative pressure, sucking up material from the substrate as it goes (Fig. 6.10). This vacuum-like action is common in fish that feed on animals or other material associated with a substratum.

The carp is thus a forager, not a pursuer of prey. It does not have the ability to swim rapidly and nor does it have teeth to grasp something large and flapping, so it rarely eats other fish (Sibbing 1988). We have at times observed common carp hunting and herding fish prey, but this is unusual. The normal behaviour for carp is using its pectoral fins to move slowly through shallow waters with soft substrates, sucking up food as it goes. Typically, a carp will move in a straight line, rotating its mouth from side to side so as to cover a broad area as it goes (Gidmark *et al.* 2012). This action has been called 'mumbling' (Cahn 1929) or 'roiling' (Wiltshire *et al.* 1962a) and macrophytes can be uprooted during the process.

The carp has two conspicuous pairs of sensory barbels or 'whiskers': one short, thin pair on the middle of the upper lip and one longer and thicker pair near the corner of its mouth (Fig. 6.11) (Farag *et al.* 2014). These sensitive tactile and taste organs are used to locate food when visibility is low.

Food of carp in Australia

Although the diet of carp in Australia has received surprisingly scant attention, it differs little from that of the fish overseas. Larval common carp are visual feeders, that feed on zooplankton, including microcrustaceans (cladocerans and copepods) and rotifers (tiny animals that are ubiquitous in freshwaters) (Hall 1981; Vilizzi and Walker 1999b), and when very small, algae. When the fish is almost a juvenile, small chironomids begin to be included in the diet (King 2005). Bottom-feeding becomes more common as carp larvae grow and transition into juveniles (Vilizzi and Walker 1999b; King 2005).

As with many species, the largest thing a carp larva can eat is determined mostly by size of its mouth. Very young carp larvae have a gape of only about 0.4 mm, whereas the gape of the oldest stage of carp larvae is about 1.2 mm (King 2005). Khan (2003) found that amphipods or scuds became more important in lake-dwelling common carp larvae as they grew. Juvenile carp feed much more extensively on the bottom on chironomids and other insect larvae, such as true bugs and microcrustaceans, including ostracods or pea shrimp (Hume *et al.* 1983; Vilizzi and Walker 1999b; Khan 2003; King 2005). Adults, like their overseas counterparts, are omnivores. They eat zooplankton, chironomids, aquatic insects, small snails and bivalves (often associated with macrophytes), seeds, detritus (including dead macrophytes) and, occasionally, filamentous algae (Hall 1981).

Although carp of all sizes ate approximately similar types of food, Hume *et al.* (1983) found that small (<150 mm), medium (150–400 mm) and large (>400 mm) carp overall showed differences in their diets in the lakes and river from which they were collected. Small carp tended to eat a greater proportion of microcrustaceans than did medium and large carp. Medium carp ate more benthic insects than the other two sizes of carp, and all sizes ate plant material. This comprised more than a third of the overall diet of large carp, whereas it was more like one-fifth of the diet of small and medium carp. The differences in diet with size are likely related to the gap between gill rakers: the larger the fish, the larger the gap, meaning that different-sized prey are retained or lost in the filtering process (Hall 1981). Hume *et al.* (1983) also found that diet varied by season and water body, and reflected the changes in types of food available.

*****

We have seen that, although there are minor differences in the way that carp does things in Australia versus overseas, there are few surprises to be found in the biology of carp in Australia. We do not see a completely foreign set of traits emerging in Australia – at least not yet – even with the very different Australian environment into which the common carp was thrust. It seems to have been as successful here as it has been everywhere else. As we shall see in the next chapter, the traits possessed by carp have allowed some populations to survive close to a hundred years in the wild after the species was first introduced to Australia. But then came the BEH, after which those traits, together with some mixing of genes and major flooding, combined to generate a fishy tsunami that to date has been virtually unstoppable.

7

The breaking storm

There followed a calm of about ten months. The 'storm' broke in 1963 ... (Wharton 1971, p. 8).

THE YEAR 1963 MUST HAVE BEEN A HEART-BREAKING ONE FOR ALF BUTCHER and his colleagues in the Victorian Department of Fisheries and Wildlife. They thought they had succeeded in poisoning all of the Boolarra carp. They so nearly had. But ultimately they were defeated, although not by the sheer number of dams in which carp were stocked. After all, they poisoned almost 1,400 dams. They were defeated by finding out too late that at least two lakes – Yallourn Dam in Gippsland and Lake Hawthorn near Mildura – had been stocked some years previously with carp. Both were large water bodies in which carp were able to breed. And most importantly, both had connections to river systems – one to the Latrobe River in Gippsland and the other to the Murray River – which gave the carp access to vastly greater areas.

With such an enormous number of releases, some carp were bound to have escaped the carp-kill team. But with the benefit of hindsight, the evidence was there in black and white in the testimony given to the Inquiry by both Colin Eric Pratt and Geoffrey William Scott that at least one Gippsland dam and Lake Hawthorn were places of serious interest. It is easy for us, comfortable in our academic armchairs 60 years on with all the facts at our fingertips, to claim that, had Scott been taken at his word, perhaps the most devastating of all the carp releases may have been prevented.

This chapter describes the circumstances whereby carp were discovered to have evaded the carp-kill program's poisoning campaign and escaped into two major river systems. We do our best to follow in time and space carp's rapid spread throughout the Murray–Darling Basin. We have made use of newspaper reports of the time, articles in periodicals, first-hand accounts and the results of fishing competitions to track the spread. We use a couple of examples of how people responded to the spread of Boolarra carp: first, commercial fishers and second, the Canberra community as carp converged on and then invaded Lake Burley Griffin. Commercial fishers were quick to capitalise on the opportunities offered by the carp invasion, and Canberrans recognised the risk to their iconic water body and the upstream trout fishery. The chapter also considers the possible reasons why the spread was so rapid, what effect the existing populations and strains of carp may have had on the success of the Boolarra carp, and how the climate and environment conspired in the process of dispersal. While doing so, we consider why, after carp had been present in Victoria and the Murray–Darling Basin for perhaps close to a century, the Boolarra carp were so successful.

THE BREAKING STORM

Much of what follows relies on Jim Wharton's account of the unfolding story of the spread of carp in the early 1960s in Victoria (Wharton 1971). At the close of 1962, following the noxious species legislation rushed through Victorian Parliament and the vigorous carp-kill program that ensued, the eradication of carp from Victoria appeared to have been a complete success. Previously, during the State Development Inquiry into the Introduction of European Carp into Victorian Waters, Colin Eric Pratt was asked point blank whether Boolarra Fish Farms had sold any carp prior to this. He answered that the company had sold 10,000 (Wiltshire *et al.* 1962b, p. 187). As we saw in Chapter 5, records had not been kept on who was sold the fish, and furthermore Pratt stated that the carp could have been sold again by the first purchasers.

The Inquiry was concluded, the recommendation went against the carp farmers and the species was declared noxious in the same year. The carp-kill campaign, as we have seen, poisoned almost 1,400 dams and several hundred re-poisonings suggest the campaign achieved total kill. But in September 1963, the hopes of a successful eradication of carp from Victoria were dashed with the discovery of carp in Serpentine Creek, a tributary of Tanjil River near Moe – about 40 km north of Boolarra – and in the Yallourn Storage Dam, which is fed by Tanjil River. The dam level was lowered and the water poisoned a month later. The chance of eradicating carp from such a large water body was slight, but the Department of Fisheries and Wildlife nevertheless tried. It copped considerable criticism from the general public, and questions were asked in the Victorian Parliament about the safeguarding of trout in nearby streams.

Only a few trout, some eels and a solitary carp succumbed to the poisoning of Yallourn Dam. Early 1964 brought more devastating news: many small carp were found in the lake at Yallourn, indicating that carp were not only still there, but had been breeding. Shortly after, more carp were found upstream of the dam, but by then any thought of more poisoning was out of the question. The situation went from bad to worse. It was discovered that many farm and forestry dams in the Latrobe Valley contained carp, despite many owners being unaware of this. Then carp turned up in Hazelwood Dam at Morwell, Lake Guthridge at Sale to the east and Moondarra Reservoir to the north. It seems that carp had been stocked or had found their way into the Latrobe River, which meant that their spread throughout a large section of south-eastern Victoria was inevitable. By early 1965, evidence was unequivocal that carp had spawned again in Yallourn Dam, they were in huge numbers and were spreading fast.

Back in 1961, the Department of Fisheries and Wildlife attempted to contain the Boolarra Fish Farms operations to their ponds at Boolarra and a State Electricity Commission (SEC) storage dam near Morwell, which was most likely the top pond of the series of Hazelwood cooling ponds (Department of Fisheries and Wildlife 1952–1961; Wiltshire *et al.* 1962b)

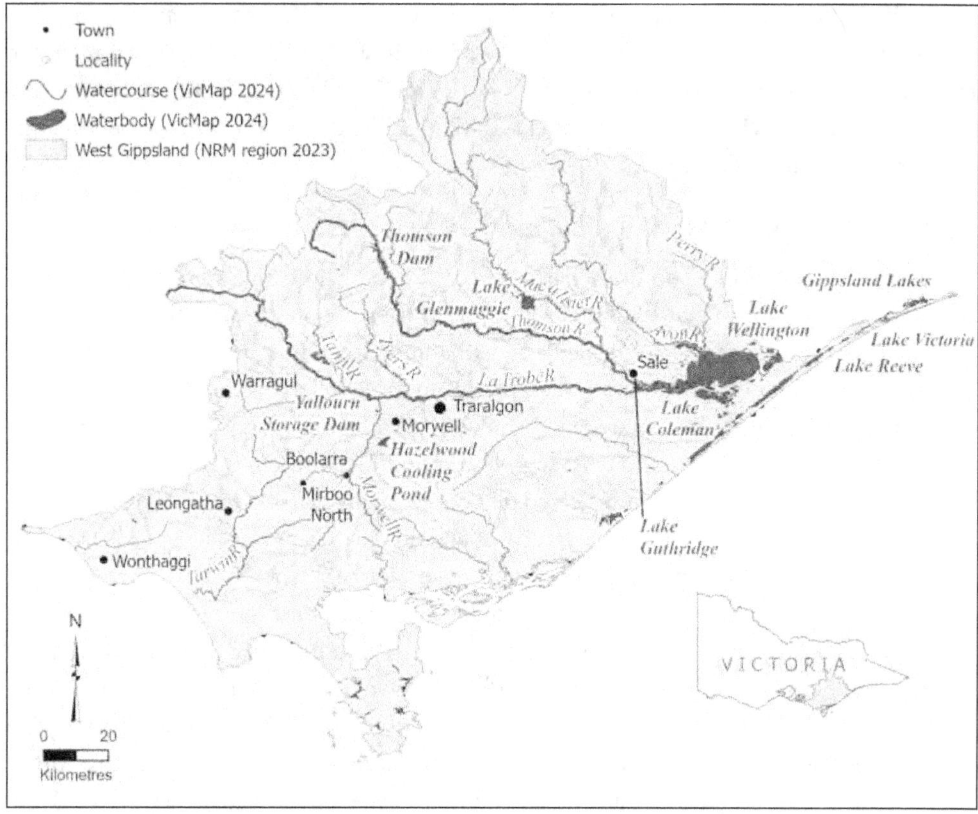

Fig. 7.1. The Latrobe River valley, showing some of the major lakes, impoundments, tributaries and towns that feature in the spread of common carp from Boolarra Fish Farms throughout the Gippsland region of Victoria. Source: SPAN, Charles Sturt University.

(Fig. 7.1). The size, location and proximity to Bennett's Creek mentioned in Butcher's report, makes the Hazelwood pond the mostly likely SEC water body in which the Boolarra Fish Farms proposed to grow out carp. It is highly improbable that, without assistance, carp escaped the Hazelwood pond into the Morwell River and made their way upstream into the Latrobe River and from there into Yallourn Dam. It is much more likely that carp had assistance to move from the Hazelwood pond to Yallourn Dam. Assistance by whom? We will probably never know.

In 1967 (a fateful year for carp invasion, as it turned out), the Victorian Piscatorial Council, SEC, Latrobe Valley Water and Victorian Department of Fisheries and Wildlife met to talk carp (Bowerman 1975). Various trapping and netting methods were tried, but proved unsuccessful. Then a distraction in the north of the state suddenly redirected everyone's attention. Within a few years, carp had spread north, south, east and west, throughout much of Gippsland. By 1971, it had been sighted in sections of the Snowy River, Stony and Boggy creeks, Tambo, Perry, Avon, Macalister, Thompson, Tyers, Morwell and Tanjil rivers and Billies and Middle creeks (tributaries of Morwell River) and, notably, Gippsland Lakes. It was seemingly unstoppable.

However devastating the news was of the escape and spread of carp into the Latrobe Valley in 1963 and its spread throughout Gippsland and rivers, a few years later, worse – much worse – was to come.

LAKE HAWTHORN

In a report written on 8 May 1961, Alf Butcher described a visit to Boolarra and noted the existence of not just one, but two companies wishing to farm the 'European carp': Boolarra Fish Farms and Geoff Scott & Co. Pty Ltd:

> *I understand that Scott has been negotiating with the First Mildura Irrigation Trust for the use of the irrigation and drainage lakes and sufficient adjacent land for a hatchery and ponds. These lakes are just outside Mildura (Department of Fisheries and Wildlife 1952–1961, 8 May 1961).*

The rest of the report deals with the Boolarra Fish Farms and there is no further mention of Geoff Scott & Co. But Scott features prominently in the 1962 Victorian State Development Inquiry into European carp farming that followed this visit, and is pivotal to the story of the escape of carp into the Murray.

We'll let Jim Wharton describe the next stage in the carp saga in his own words before going into the details:

> *While still suffering shock of the loss of the Latrobe River system to European carp we suffered yet another bitter blow in August 1967 when Fisheries and Wildlife Officer Bill Kelly reported from Mildura that Lake Hawthorn [Fig. 7.2], a salty irrigation drainage lake which has a short channel connecting it to the Murray River was 'full of carp' and of all sizes up to 10 lb. The fish had obviously been placed in the lake some years before and had bred several times. (At this point it is interesting to note that a person associated with the carp farm at Boolarra applied to the Department in 1962 for permission to use Lake Hawthorn as an experimental water for European carp. Permission was not given) (Wharton 1971, p. 8)*

The 'person' to whom Wharton referred was Geoffrey William Scott, in 1962 a resident of Boolarra with his eye on a business opportunity in the north-west of the state, where waters

Fig. 7.2. Lake Hawthorn, near Mildura. Source: Paul Humphries.

are warmer and the prospects for carp farming promising. In testimony to the Victorian State Development Committee in February 1962, Scott described himself as a 'general contractor residing at Boolarra. My business covers bulldozing, clearing, logging, road-making and dam construction' (Wiltshire *et al.* 1962b, p. 16):

> *My company has been engaged in those occupations for approximately 10–12 years, and has carried out work all over Victoria. When I refer to 'all over Victoria', my company has in recent months moved over to the Mildura district to develop land in that area should we get the 'green light' for this project. My company has leased Lake Hawthorn at Mildura from the First Mildura Irrigation Trust, because of the warmer climate in that district.*
> *On behalf of Geoff Scott Pty. Ltd I therefore wish to give evidence before this State Development Committee relating to King carp in connection with commercial fish farming. I refer to King carp (*Cyprinus carpio*). This is an industry which, if allowed to develop freely and without restriction, must ultimately be of great benefit not only to Victoria, but to Australia as a whole (Wiltshire* et al. *1962b, p. 16).*

Scott quoted from the Australian Encyclopaedia (Volume 4) about the widespread and early introduction of various species of carp to Australia, including crucian carp and goldfish. He argued that risks to the environment were exaggerated and minimal, carp 'will die or be eaten by other predators without husbandry' (p. 16), they are 'vegetarians' (p. 17), 'will not be detrimental to our wild duck and other water birds' (p. 21) and, what's more, they will provide good sport fishing for farmers in their own farm dams (p. 17). Scott argued his case at length:

> *My company leases Hawthorn Lake, near Mildura, from the First Mildura Irrigation Trust. It does not seem consistent that the Fisheries and Wildlife Department should direct that we cannot stock Lake Hawthorn – which as I have said we lease for fish farming purposes – or any other Government water because they fear that King carp (*Cyprinus carpio*) may escape and become a pest. What difference would it make if they did escape? They have already been here for 100 years! Have they yet to become a pest? No Sir, they are almost non-existent, although present stocks have originated from Australian waters (Wiltshire* et al. *1962b, pp. 19–20).*

Scott continued:

> *The fact that I stand before this Committee today is proof of my interest and confidence in the demand for King carp (*Cyprinus carpio*), coupled with the time and money expended by me in investigating by aeroplane and motor car, sites suitable for the building of dams by bulldozers for this project (Wiltshire* et al. *1962b, p. 21).*

He, like Pratt, had already invested heavily in the venture. It is uncertain from Scott's testimony to the State Development Inquiry whether he had been undertaking breeding and rearing trials in ponds in a Gippsland location separate from those at Boolarra, but it seems clear that he had carp in a pond he owned at Moe (Wiltshire *et al.* 1962b, p. 166). Scott said he had got the fish from Boolarra. In answer to a question from Vice-Chair of the Committee, George Schintler, as to whether Scott had sold live carp from his pond for breeding purposes, Scott gave an affirmative and said that he had obtained his fish from Boolarra Fish Farms but had not distributed them widely (Wiltshire *et al.* 1962b, p. 22). The Inquiry questioning for a while went in different directions but it later returned to the issue of stocking and rearing of carp.

During Scott's testimony to the Inquiry in February 1962, Lake Hawthorn was mentioned several more times, including during his recall by the committee. Scott showed photographs of Lake Hawthorn and pointed out clear signs of 'aquatic plant destruction', claiming, however, that 'there are no known *Cyprinus carpio* in this lake' and that other fish must be responsible (Wiltshire *et al.* 1962b, p. 166):

> *Chairman (C): Mr Scott, to the best of your knowledge and belief there are no*
> *carp in Lake Hawthorn?*
> *Scott (S): So I have been told there are none there.*
> *C: To the best of your knowledge?*
> *S: Yes. There may be* Carassius carassius *[crucian carp].*
> *C: But not* Cyprinus carpio?
> *S: I do not know whether there is or not; I have not dragged it (Wiltshire* et al.
> *1962b, pp. 175–176).*

There followed some discussion about Lake Hawthorn and the cause of the water plants being 'torn out'. Scott argued it was done by fish, although what type of fish was not ascertained, whereas the Chairman suggested that the cause might be boats and water skiers. Then the discussion returned to carp and Scott's knowledge of where carp had been sold and distributed. But the talks became circular and we, and presumably the Committee of Inquiry, were left in some doubt as to whether carp were in fact in Lake Hawthorn. From what we know now, we can speculate that they were already in the lake when the Inquiry was being held. We can be fairly confident that carp made it into Lake Hawthorn – and perhaps had access to the canal that connected it with the Murray River – as early as 1961 or 1962.

This speculation has support from Officer Kelly's observation that in 1967 the lake held carp of 'all sizes up to 10 lb' (4.5 kg). Assuming that Officer Kelly was correct and that one or more fish were actually this big, Vilizzi and Walker's models of carp in the Murray River (Vilizzi and Walker 1999a) indicate they were somewhere from 4–10 years of age. It is possible that a 10 lb (4.5 kg) fish was an adult when released into Lake Hawthorn, in which case we would have no idea how long the fish had been there. But if it was the progeny of released fish and hatched in the first year that fish were in Lake Hawthorn, it would mean that the latest year that carp were stocked in the lake was 1963 and the earliest was 1957. If we average the two, then the year is 1960.

Two final exchanges between the Committee Chairman and Scott are worth mentioning. Scott was adamant, that like overstocked sheep or cattle in a paddock, if there were too many carp in a pond, dam or lake then one would simply remove a portion of them, thus taking pressure off the system and improving the condition of the water body. This would be achieved by a 'fishmaster' appointed to each area, whose job was to remove excess fish when necessary. Scott's arguments were about lakes and ponds. The Committee was not satisfied with Scott's analogy and followed up with what we now see as insightful and prophetic questioning:

> C: *You are familiar with the River Murray system.*
> S: *Yes.*
> C: *How would you net them [carp] there?*
> S: *You could not net them.*
> C: *You could not take out half of the fish?*
> S: *No. But you could use other methods, possibly this new electronic method.*
> C: *I believe that has not been a success?*
> S: *It does not have to be 100 per cent successful. If as has been explained to me the fish can be controlled the same as any other animal, it could be used to stop overstocking, which is the main problem.*

Finally, Chairman Wiltshire returned to the reply to a letter Scott had written to the curator of fish at the Australian Museum, G.P. Whitley, seeking information about introductions of carp to Australia:

> C: *In his letter to you Mr Gilbert P. Whitley has said '... I have only just heard of its [carp's] appearance in the Mildura region'. Was he referring to any statement made by you?*
> S: *I do not know what reference he has there at all. It came as a surprise to me, I don't mind telling you, he heard of its appearance. I thought I was the only one who made that enquiry.*
> C: *Where was it in the Mildura region to your knowledge?*
> S: *I do not know. I do not know any more than you (Wiltshire et al. 1962b, p. 178).*

And they all swam wild

The tragedy of carp getting away from fisheries authorities in Gippsland cannot be downplayed. But its release in Lake Hawthorn and subsequent escape into the Murray–Darling Basin (what we termed the Boolarra Event Horizon or BEH) was a disaster, the effects of which we are still feeling to this day. As we have described, there is circumstantial evidence that carp had been in Lake Hawthorn for several years before it was discovered by Officer Kelly in 1967 and that it had already made its way into the Murray downstream of the canal that linked the two water bodies (Fig. 7.3). In fact, carp likely entered the Murray River perhaps by 1964 or even earlier, because Lake Hawthorn connects to the Murray River when flows are more than 90,000 ML/ day (Forsyth *et al.* 2013). This was the case in November 1960 and again in November 1964 (Fig. 7.4). There was a fairly long dry period during the 1960s, and indeed 1964 was a bit of an aberration. The 1960s drought broke with a vengeance in 1973, with big floods from 1973–1975, which we will come back to later in the chapter. For the moment, we will detail the spread of carp, for the most part chronologically.

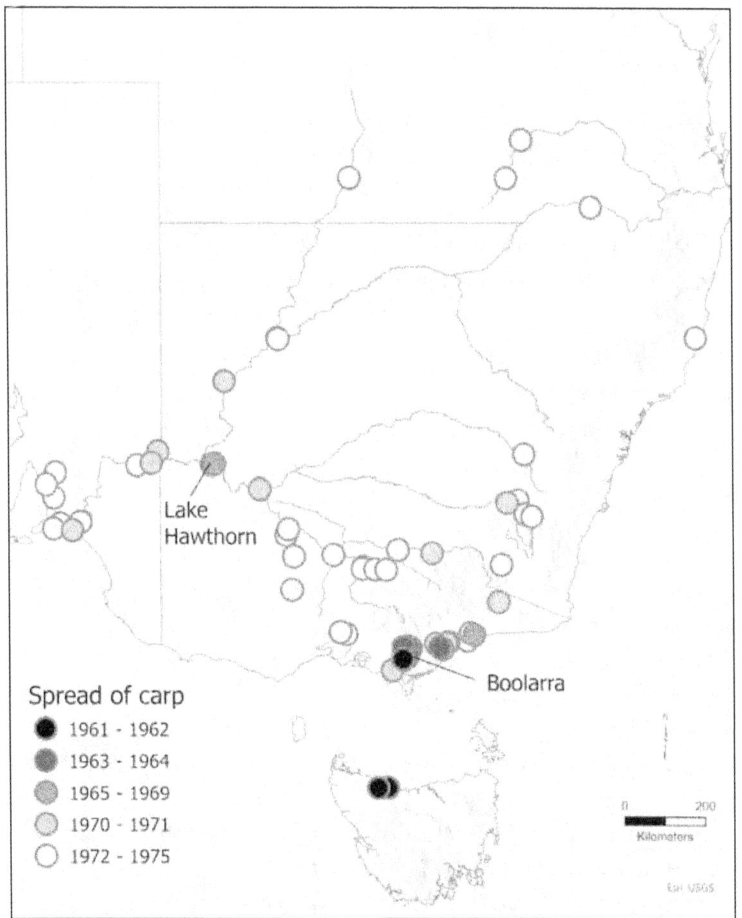

Fig. 7.3. The chronological spread of common carp throughout south-eastern Australia, following its introduction to Boolarra Fish Farms. Source SPAN, Charles Sturt University.

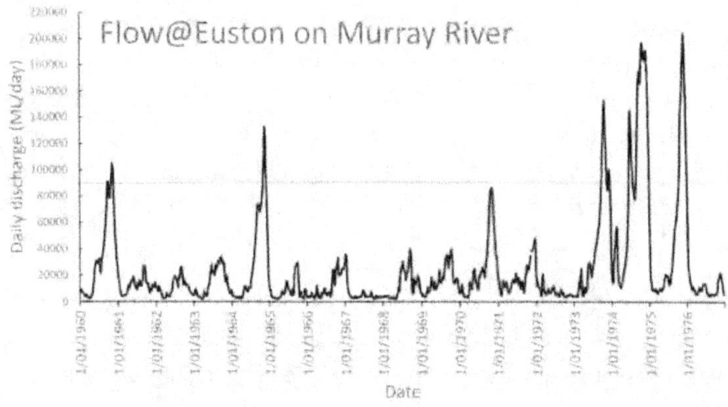

Fig. 7.4. Flows in the Murray River at the Euston gauge, 1960–1976. Dashed line indicates flow at which Lake Hawthorn connects with the Murray River. Source: Data from Murray–Darling Basin Authority.

1964–1969

A letter to the *Canberra Times* by a Mr Taylor of Yarralumla in September 1965 claimed:

> *... they [common carp] are certainly lurking in the billabongs of the Murray system, eager to pit their wits against any angler with the patience, the skill and the determination to outwit them (20 September 1965, p. 2).*

However, there is no hard evidence that Boolarra carp had spread widely in the Murray system as early as 1965. Taylor clearly knew his crucian carp and goldfish from his common carp, as he gave fishing correspondent Hec Horsburgh a lesson in fish taxonomy and origins, but this remains anecdotal evidence at best. It would mean either a rapid dispersal from Lake Hawthorn or recognition of the existence of previously known populations, like the Yanco carp. This concern was echoed in an article in the *Australian Fisheries Newsletter* at the time, which asked anglers not to use small carp as live bait and thus not further spread carp around:

> *The Fisheries and Wildlife Department of Victoria is concerned at reports that European carp are breeding prolifically in Yallourn Storage Dam, the pondage at Hazlewood, and other Gippsland waters, including Lake Gutheridge* [sic] *at Sale. The Director of the Fisheries and Wildlife Department, Mr. A. Dunbavin Butcher, said that his Department's efforts to keep European carp out of waters in the Latrobe Valley had failed.*
>
> *Mr. Butcher said that this was matter of great disappointment. He hoped that carp would not replace trout, but only time would tell.*
>
> *The European carp (Cyprinus carpio) can be recognised by the feelers or barbels on each side of its mouth. In 1962 this fish was proclaimed a noxious fish in Victoria under the Fisheries Act. Anyone who had these carp in his dam had to notify the Fisheries and Wildlife Department. In all, nearly 1,400 dams were treated and the carp eradicated by Departmental officers. In 10 of the 11 areas of Victoria's treated, the carp eradication programme had been successful, but in the Latrobe Valley many dams had been stocked with carp without the knowledge of property owners. These dams were therefore not reported to the Department, and so did not receive treatment ...*
>
> *Whether the carp would be prevented from spreading from Gippsland to other areas was a question which could be influenced greatly by the co-operation of sport fishermen – one important factor in limiting such a movement is that anglers should not transport small bait fish from one place to another (*Australian Fisheries Newsletter *1965, p. 17).*

There is no mention of carp being in the Murray at that time, which is a little surprising since it seems that commercial fishers had been catching it in various Murray–Darling Basin rivers for years. Nevertheless, the spread of carp became a common subject in this publication over the ensuing years.

John Lake's 1967 book *Freshwater Fish of the Murray–Darling System: The Native and Introduced Fish Species* has little to say about common carp, and was clearly published just before the Lake Hawthorn revelations. If carp had been widespread in the Murray, we might have expected more than this:

> *At present this species is not a nuisance in New South Wales but recently there were fears in Victoria that its widespread use in farm dams might lead to trouble and it was therefore declared noxious. It would be wise to see that its further spread is prevented where possible (Lake 1967).*

Alan Weatherley and John Lake also published a chapter on 'Introduced fish species in Australian inland waters' in the same year, and although 'European carp' received little consideration, its existence in the Murray River was mentioned. 'It is rare in main rivers, but present though not abundant in irrigation channels in New South Wales' (Weatherley and Lake 1967, p. 225). Again, one would expect that if carp was abundant and/or spreading at the time, this would have been mentioned – especially considering the focus of the chapter. Weatherley and Lake may have actually been referring to the Yanco strain of carp, because their map clearly shows carp within a small area associated with the Murrumbidgee River, which is consistent with the later results of Shearer and Mulley (1978). What adds to the confusion is that Weatherley and Lake (1967) stated that carp was found 'in the Murray River' (p. 225) but with a map that showed the Murrumbidgee River, while Shearer and Mulley (1978) stated that in their study they sought to determine if the Boolarra fish spreading through the Murray–Darling Basin was the same strain of fish that had been in Prospect 'and Murrumbidgee Irrigation Area which were described by Weatherley and Lake (1967)' (Shearer and Mulley 1978, p. 551). We can only assume that Shearer and Mulley interpreted Weatherley and Lake's distribution diagram in the same way that we have.

1970–1971

The first consistent and credible evidence that carp was in all likelihood spreading upstream and downstream from Lake Hawthorn in the Murray came in 1970 (Fig. 7.3). It cannot be ruled out that some of the following instances were of pre-existing populations of the species, but there are multiple lines of evidence that carp was being noted where it previously had not. There were reports that carp had reached Chowilla Dam in January 1970 (Bowerman 1975), about 250 river kilometres downstream of Lake Hawthorn. Then there was a report of an 11 lb fish (5 kg) caught near Lock 6, east of Renmark (Australian Fisheries Newsletter 1970) about the same location (perhaps it was the same fish). Further downstream in South Australia, a 10 lb (4.5 kg) carp was caught at Pelican Point on the Coorong by a professional fisherman (*Victor Harbour Times*, 13 November 1970, p. 5) – 800 river km or so downstream from Lake Hawthorn. The same year, there were less definitive reports that 'numerous fish were being caught in the upper Murray River by November' (Bowerman 1975, p. 5). If the reports in the upper Murray were true, even if the area referred to was only Albury–Wodonga and if the source of the carp was Lake Hawthorn, that is impressive movement in only a few years (Dean Gilligan's ageing of carp in Lake Hume, however, indicated that carp may not have made it above the dam wall until 1977, although those results are based on surviving fish caught in

2004 (Gilligan and Rayner 2007)). The *Victor Harbour Times*, citing the evidence of fishers, reported the proliferation of carp in the lower Murray by April 1971: 'The [South Australian] Department of Fisheries and Fauna Conservation now have sufficient reports of catches of European Carp to confirm that they had spread throughout the River Murray and Lakes System' and the fact was so established that the Department Director said that no more reports were necessary (*Victor Harbour Times*, 2 April 1971, p. 1).

Wharton claimed that by 1971 carp occupied the Murray westward from about Boundary Bend (according to his map) in Victoria (Wharton 1971). From the above reports, carp occurred throughout much of the South Australian Murray too, and was beginning to make its way up the Darling and Murrumbidgee rivers (The Riverlander 1974). Although 'carp' of indeterminate species had been part of commercial fisheries' catches for

Fig. 7.5. Newspaper advertisement for 'European carp' from the Murray River. Source: Redrawn from *Australian Jewish News*, 19 February 1971, p. 8.

many years prior to 1971, carp had become abundant enough by that time for the species to become commercially viable in its own right. So it was around this time that commercial fishing of carp took off in earnest. Indeed, European carp 'direct from the Murray River' were advertised in Melbourne newspapers (Fig. 7.5).

1970–1974

Fortunately, we have records of commercial catches for New South Wales and South Australia for at least part of the expansion of carp (Figs 7.3, 7.6 and 7.7). The best data are from New South Wales (Reid *et al.* 1997), for which there are catch returns from individual catchments (Pease and Grinberg 1995; Forsyth *et al.* 2013). The data from South Australia are limited (Pillar 1976). Apparently there were five Victorian commercial carp fishers, two who used electrofishers and three who used nets in lakes in the north of the state (Bowerman 1975). Unfortunately, we don't have data from those catches. While some of the carp caught were destined for human consumption, large quantities were sold for the production of pet food or as rock lobster bait (Bowerman 1975; Davis 1975).

Commercial catch data show that carp first became a prominent part of the commercial fishery in the mid-Murray from 1970 (although there were low catches of 'carp' from at least the 1950s), and catches increased dramatically (Fig. 7.7). This fits with the first newspaper advertisements of carp for sale. The catch of carp in the Darling and Murrumbidgee rivers was a little more delayed: there were small catches in 1970, but they increased substantially from 1973. Interestingly, the Lachlan River and lower Murray regions had variable catches of carp until the mid-1970s, after which catches increased dramatically, more in the latter than the former. By 1973, the price of carp was so low – presumably because there was an oversupply, despite high red meat prices – that the return of 5–8 cents/lb was not worth the effort for professional fishers (*Canberra Times*, 7 April 1973, p. 9). A couple of years later, a carp could fetch 77 cents/kg at the Melbourne Fish Market but routinely the price was 8–25 cents/kg (Bowerman 1975).

Fig. 7.6. Catching European carp at Koondrook for processing into pet food, 1975. Source: Terry Rowe, National Archives of Australia: A6135, K31/10/75/7.

Another source of information on when carp turned up in rivers, or at least became prominent, is recreational fishing competitions. The results of some angling clubs were published in newspapers that have been digitised and/or scanned (e.g. South Australian Australian Anglers Association, https://anglersa.org.au/past.html). The Victor Harbour Angling Club held its annual freshwater fishing competition in the lower Murray, usually between about Morgan and Mannum, at varying times of the year, but mostly in late spring or summer. Fishing

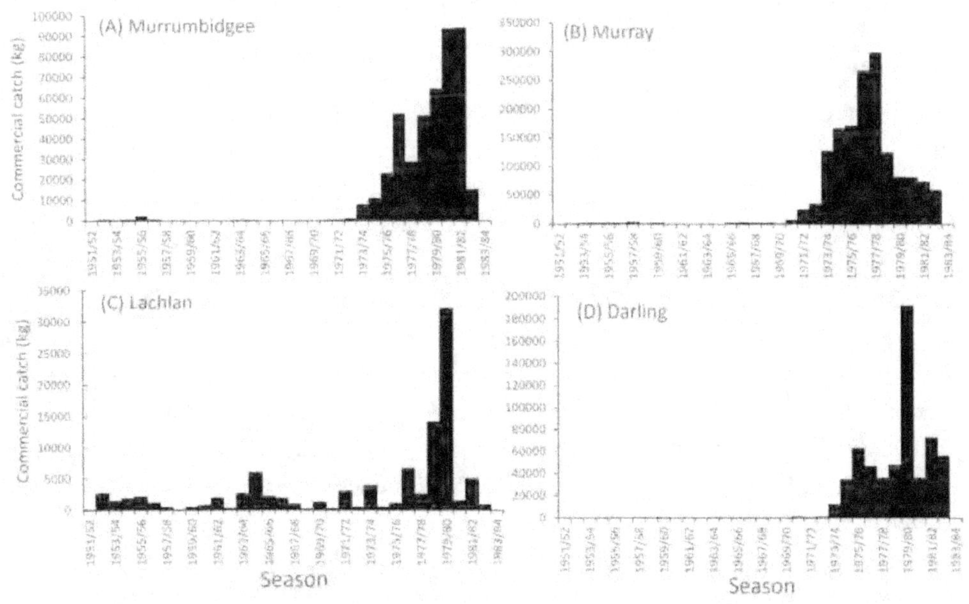

Fig. 7.7. Annual commercial catches of common carp in New South Wales by catchment, 1950–1984. Source: NSW DPI Fisheries, with permission.

competition results for 1966–1973 listed callop (golden perch) and catfish (*Tandanus tandanus*) as the dominant species caught, with the odd Murray cod and redfin (European perch) in the mix, but there was no mention of carp during that period (*Victor Harbour Times*, 1966–1973). That changed dramatically in February 1974. That month, Miss Joy Jaensch, a member of the Victor Harbour Angling Club, won the SA State Women's Freshwater Championship, and 'it was disappointing to see the large number of European carp landed' (*Victor Harbour Times*, 13 February 1974, p. 1). By November 1974, although Victor Harbour fishers caught good-sized fish they were not the ones normally favoured, as 'the European carp were in vast numbers' (*Victor Harbour Times*, 13 November 1974, p. 2). In April 1975 at Swan Reach on the Murray River, most of the 309 lb 7 oz (140 kg) catch during the fishing competition was carp (*Victor Harbour Times*, 16 April 1975, p. 6). Similar results were reported for the South Australian Australian Angling Association (SAAAA) annual freshwater competition from 1975 onwards (https://anglersa.org. au/past.html).

By 1974, carp had made its way into the upper reaches of the Yarra River near Melbourne (Fisherman's Forum 1974), which concerned those who protected the quality of Melbourne's water supplies (*Canberra Times*, 12 June 1974, p. 8). It had also spread considerable distances upstream in the Murray and Murrumbidgee rivers. Reports had come in that carp was upstream of Yarrawonga in the Murray, and was in the Loddon River and the Goulburn River as far upstream as Shepparton (Fisherman's Forum 1974). By November 1974 it had reached the spillway of Burrinjuck Dam on the Murrumbidgee River (*Canberra Times*, 22 November 1974, p. 16). It was being held back by the dam wall, but there was concern that this containment would not last long if fishers used small carp as live bait and thus transferred the species into the dam itself. Reports from recreational fishers lower down the Murrumbidgee and the Murray were dire: carp dominated catches to such an extent that few other fish were caught.

1975 onwards

By 1975, the carp invasion of the Murray–Darling Basin was all but complete. Probably a decade after the carp's escape from Lake Hawthorn, its numbers had skyrocketed along much of the length of the New South Wales and South Australian Murray River. It was in the Broken, Campaspe, Loddon and Avoca rivers and their tributaries and in Kerang Lakes in Victoria (Bowerman 1975). In South Australia, it was in the Lower Lakes (including their tributaries), Lake Bonney, Currency Creek, Finnis River, the Torrens River near Adelaide and the Light River on the Adelaide Plain (Buckmaster 1975). Carp was in the New South Wales section of the Paroo River and there was a risk that it would get into the Cooper Creek system. In a letter to Fred Reynolds (who was appointed in 1974 to study native and introduced fish in South Australia) from minutes of a meeting between Don Buckmaster (Arthur Rylah Institute in Melbourne), Reynolds and Karl Shearer (NSW Fisheries), it was confirmed that in South Australia, carp:

> *is the second most abundant fish in the lower sections of the [Murray] river – bony bream (*Fluvialosa richardsoni*) being the most common. During the flooding in 1974 the barrages at the mouth of the Murray were opened to allow excess*

freshwater to escape and thousands of Common carp were flushed to sea and survived for varying periods in the low salinity water near the Murray mouth (Buckmaster 1975).

Commercial fishing for carp was in full swing. It was peaking or close to peaking in many jurisdictions in South Australia, Victoria and New South Wales in 1976 (Pillar 1976; Reid *et al.* 1997; Forsyth *et al.* 2013). In Victoria alone, carp was the fourth largest commercial fishery, and there was a suggestion at the time that amateur fishers could assist with removal of carp, and sell them (Australian Fisheries Newsletter 1976b). Apparently, a contingent of enterprising Stanhope Apex Club members in northern Victoria had collected 4,000 kg of carp using pitchforks! Mr Heighway, Executive Officer of the Victorian Farmers' Union's Commercial Fishermen's Section, responded negatively to this suggestion, opining that amateurs 'had other avenues of earning income' (Australian Fisheries Newsletter 1976b, p. 27).

There was genuine fear that it was only a matter of time before carp got into Queensland. A year later, those fears were realised. There were confirmed reports in 1976 of carp in the Queensland section of the Paroo River, and in the Balonne and Dumaresq rivers (*Canberra Times*, 22 April 1976, p. 3). Over the following years, carp continued its inexorable spread, not just in but also outside the Murray–Darling Basin. Carp is in the Upper Condamine River (Frawley *et al.* 2012); various wetlands in Perth, the upper Swan River, and anecdotally in Mandurah, Bridgetown and Harvey in Western Australia (Breheny 1996); Leigh Creek Dam in the Lake Eyre Basin (Brown 1996); coastal catchments of the Hunter, Hawkesbury, Central Coast and Shoalhaven regions of New South Wales (Faragher and Lintermans 1997); the Albert–Logan River system in Queensland (Norris *et al.* 2011); and most notably reached Tasmania (again) in the mid-1990s (Diggle *et al.* 2012; Stuart *et al.* 2021). In the last decade or two, koi has also found its way out of the fishponds or tanks of enthusiasts into many waterways, including the Richmond, Bellinger, Hastings, Wallamba, Towamba and Macleay catchments of coastal New South Wales (Graham *et al.* 2005; Haynes *et al.* 2010).

LAKE BURLEY GRIFFIN

By 1976, carp was in all eastern mainland Australian states, the Australian Capital Territory and almost every major river in the Murray–Darling Basin. One particularly prominent water body, Canberra's Lake Burley Griffin, an icon of the nation's Capital Territory, had been free of the carp scourge until the mid-1970s. Carp's invasion of the lake was inevitable. The response of authorities and the general public is a case study of fear, uncertainty and responses to rapid unwanted change in the lead-up to and following a species invasion.

Even though the Yanco strain of carp had been in the Murrumbidgee system for probably 90 years, the newcomer on the block – the Boolarra carp – had been steadily making its way up the Murrumbidgee since about 1970. Burrinjuck Dam had impeded its movement upstream, probably delaying its progress for several years. Indeed, Shearer and Mulley (1978) claimed that the upstream movement of carp had been prevented by Yarrawonga Weir on the Murray from 1972–1974, Menindee Weir on the Darling from 1971–1974 and Wyangala Dam on the Lachlan and Burrinjuck Dam on the Murrumbidgee since about 1977. The authors seemed unaware that the carp reported to be downstream of Burrinjuck in 1974 had by 1976 made its way above the dam, into the Molonglo River and from there into Lake Burley Griffin (*Canberra Times*, 13 February 1976, p. 16). There were dozens of carp in the lake in February of that year (*Canberra Times*, 27 February 1976, p. 22). By the following year, carp was in much larger numbers which included 1- and 2-year-old fish, weighing 2–4 kg. 'These fish are in such numbers in some parts of the lake that it is possible to wade among them, and some fishermen have scooped them up with landing nets' (*Canberra Times*, 21 October 1977, p. 17).

Fig. 7.8. Advertisement for the 1978 Lake Burley Griffin carp fishing competition. Source: *Canberra Times*, 10 March 1978, p. 6.

The concern about the movement of carp up the Murrumbidgee River and from there into the highlands and the Snowy Mountains included the fish's possible effects on the aquatic environment and the trout fishery (e.g. *Canberra Times*, 22 November 1974, p. 16). This focus was much repeated in newspapers, admittedly most commonly by anglers and angling journalists. But there was also general concern for native fish conservation and water quality in Lake Burley Griffin and elsewhere, which was reflected in Australian Federal Parliament question time on multiple occasions from 1975 (e.g. *Hansard* 14 May, 5 September 1975; 17 March, 23 March, 4 May, 23 September 1976; 17 August, 8 September 1977; 9 March, 11 October 1978).

Carp had become so abundant in Lake Burley Griffin by 1978 that a carp fishing competition was held as part of Canberra Day celebrations in March (Fig. 7.8) in an attempt to harness the power of citizen science to collect biological data (*Canberra Times*, 10 March 1978, p. 6). There was $30,000 up for grabs, more than 200 tagged carp worth $100 each and a specially tagged carp that would have netted the catcher $10,000. This particularly valuable carp had been released in front of Old Parliament House! Tens of thousands of carp were caught, and seven trucks were needed to remove the carcases (*Canberra Times*, 13 March 1978, p. 3). Although the $10,000 carp got away, information on the status of carp in the lake and much more besides was gained. The fishing competition was repeated in 1979 and for many years after, and native species were also tagged for prize-winning.

WHY DID BOOLARRA CARP SUCCEED WHEN OTHER CARP DID NOT?

It is not uncommon for invasive species to remain dormant in low numbers for many years, even decades. Invasive species in such environments have been called 'sleeper populations' (Spear *et al.* 2021), which may wake unpredictably if circumstances change. These sleeper populations may be in low numbers for long periods and their presence may be known, or in some cases unknown if they exist in such small numbers that routine monitoring or a lack of intensive sampling misses them entirely.

The carp in Australia almost certainly constituted a sleeper population, considering that at least the Prospect and Yanco strains were in existence long before 1967. But something kept them in check for many years. It is unlikely that the 'escape' of the Boolarra strain of carp triggered the existing Murray–Darling Basin carp to 'wake up'. The evidence suggests that the Boolarra carp was awake from Day 1. Why was this new strain so unlike its resident cousins, in spreading so rapidly and proliferating so fast? As a note: most articles on the spread of carp in the Murray–Darling Basin over the last few decades concluded that the Boolarra carp was imported from Europe (Wharton 1979; Brumley 1991; Brown *et al.* 2005; Haynes 2009), but first-hand accounts suggest that its origin was Prospect Reservoir fish (Peter Rogan and Russel Wurcherpfennig, pers. comm.). The likelihood is that floods, genetics and other facilitators may have combined to create a perfect storm that allowed the raging success of the Boolarra carp. But it must be kept in mind that the Boolarra carp was successful in two river basins – Latrobe and Murray–Darling – under quite different conditions.

First, the floods of 1973–1975 would have greatly helped the spread of carp from Lake Hawthorn downstream and upstream, regardless of any other factors (Brumley 1991; Koehn *et al.* 2000; Koehn 2004). Floods allow passive movement downstream and enable movement over small weirs if they are 'drowned out'. They also provide the inundated vegetation that is ideal for carp spawning, recruitment and growth. However, there were previous floods which did not seem to have the same effect on the existing carp populations pre the BEH. So, flooding is probably not the only reason for success.

Mulley and Shearer (1980) suggested that the introduction of Boolarra carp to the Murray–Darling Basin may have succeeded due to interbreeding. Since the Yanco strain could have originated in Asia and the Boolarra (and Prospect) strain in Europe, and their separation would have meant considerable genetic differentiation, interbreeding of Boolarra and Yanco fish would have resulted in the next generation having increased genetic variation. Mulley and Shearer argued that increased genetic variation could have resulted in a breakdown in previously limited and separate populations that existed prior to the BEH, which would have provided novel opportunities for adaptations to the Murray–Darling Basin environment. Alistair Brown and Andrea Brumley, employed on the Victorian Carp Program in the late 1970s–early 1980s, made similar arguments that hybrid vigour and hybridisation with goldfish might have been a contributing factor (Brown 1980; Brumley 1991).

More recently, studies by Davis and colleagues (1999) and Haynes and colleagues shed light on the genetics of carp in Australia and the history of its spread (Haynes 2009; Haynes *et al.* 2009, 2010, 2012). The studies concluded that there are four strains of carp in the Murray–Darling Basin, and Australia more generally: the Prospect, Yanco and Boolarra, all of European origin and so *Cyprinus carpio carpio*; and koi, of Asian origin and so *Cyprinus carpio haematopterus*. The presumed Boolarra carp strain is present throughout the Murray–Darling Basin but dominant in Victoria; the presumed Yanco group is also widespread but dominant in the Murrumbidgee area; a koi group is present, mostly in the south-east of the

Murray–Darling Basin and in Tasmania. Contrary to the results of previous studies, Haynes *et al.* (2009) suggested that the range of Prospect carp may have expanded during 1950s floods, prior to the BEH. Boolarra carp may have similarly surfed a flood wave, but during the 1973–1975 floods. Yanco carp may have begun to expand only after the BEH. But if Prospect carp was also more widespread than previously thought, then the opportunity for Boolarra carp to breed with several other strains was there, once it got into the Murray (Haynes 2009).

Besides the floods of the 1970s and interbreeding of Boolarra with existing strains of carp, there are several less prominent reasons why the Boolarra carp may have taken off in Murray–Darling Basin. The spread of carp may have been facilitated:

1. through what is called 'secondary invasion' (O'Loughlin and Green 2017) – a previous invader, such as tench or goldfish, may have 'softened up' environmental conditions that gave carp a leg-up in its invasion potential;
2. by degradation of the existing fish fauna – commercial and recreational fishing and other factors may have degraded native fishes to such an extent that by the time the BEH occurred the numbers of potential predators or competitors were so low that there was little resistance to this new invasion wave (Doyle 2012; Humphries 2023); and
3. changes to the riverine environment caused by water storage and river regulation – although dam-building and river regulation had been gathering steam in Murray–Darling rivers since the late 1800s, the number of dams had increased dramatically since the 1930s and created

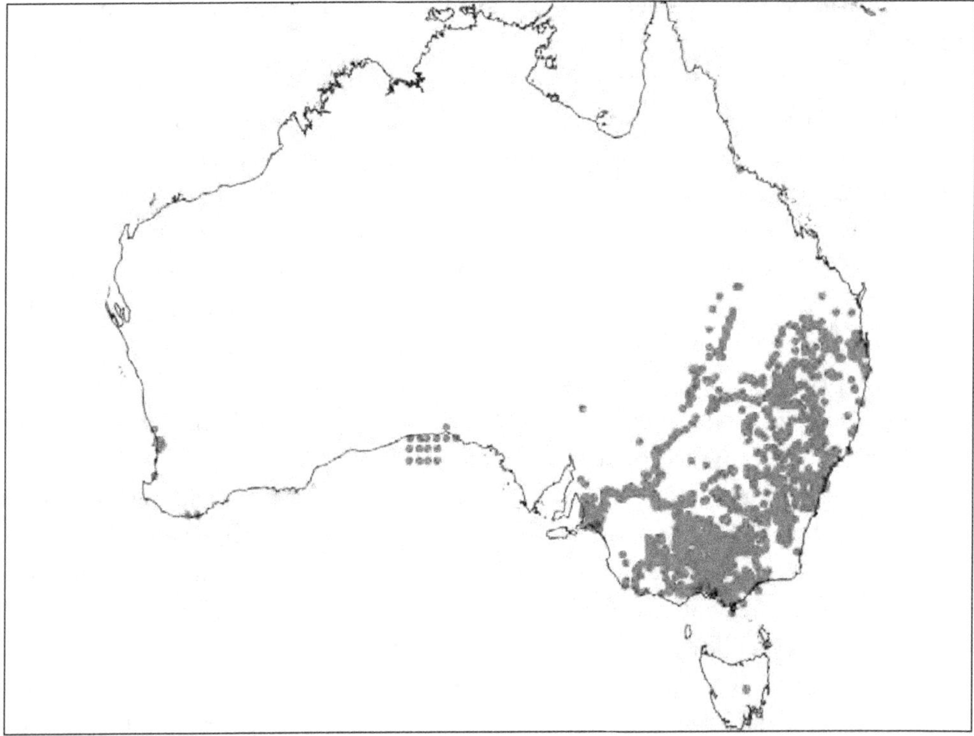

Fig. 7.9. Common carp distribution in Australia. Source: Atlas of Living Australia, December 2024.

better conditions for carp than for most native fishes (Gehrke 1997; Bunn and Arthington 2002; Koehn *et al.* 2020).

*****

Trying to put the pieces of this historical puzzle together with any certainty is virtually impossible. Many changes to the environment were happening simultaneously and rapidly, and with some of those changes, which was the cause and which the effect is not at all clear. Unlike its predecessors, the odds that the Boolarra carp would 'misbehave' (Wharton 1971) under the conditions that prevailed from the mid-1960s onwards were high. And it did. The carp genie was well and truly out of the bottle. Carp now occurs in all states and territories in Australia except the Northern Territory (Fig. 7.9). In the following chapter, we move on from the escape and proliferation of the Boolarra carp in Australia, to the impacts it has had on its new aquatic environment and on the fauna and flora which inhabits the rivers, floodplains and lakes it has invaded.

8

Impacts of carp in Australia

*Because carp were feeding on the bottom, they ate all the catfish eggs ... I haven't
seen a snail since the carp ... And that was the main food of the catties, plus all the
mussel beds, they wiped them out too. Bill Lever, professional fisherman (Frawley
et al. 2012, p. 61).*

IT IS TEMPTING TO ASSUME THAT WHEN THE CONDITION OF THE AQUATIC
environment worsens following the introduction of a non-native fish, the effects can be entirely
ascribed to the actions of that species. It is rarely the case, however, that there are pre-existing
data to compare with data collected after the introduction, and it is rarer still that an introduction
occurs in isolation from other environmental stressors (Didham *et al.* 2005). There is also often
a lag – sometimes spanning decades – from the introduction to when assessments, or at least
rigorous systematic ones, are made of the effects of the introduced species.

As with many aspects of the story of carp in Australia, the direct evidence for its environmental
impact is patchy. We peer into the murky waters of the past, hoping for a glimpse of what
actually happened. In this case, it is unclear how much harm carp has done to our waterways
relative to myriad other factors that have degraded our natural environment in the last 200
years or so. Over-harvesting of native fishes, altered flow regimes, construction of barriers
to movement, artificial lakes, thermal pollution, other non-native fish introductions, habitat
alteration, snag removal and nutrient overload all occurred prior to, at the same time as and
after the Boolarra Event Horizon (BEH).

A pamphlet from the early 2000s posed the question: 'Carp – Villains or victims?' (Fig. 8.1):

*But are carp the villains or just one of many symptoms being displayed by
our stressed rivers? Are carp a scapegoat for 200 years of inappropriate river
management, or are they one of the prime causes of degradation in our rivers?
(Murray–Darling Basin Commission 2002, p. 2).*

We are not a great fan of the term 'victims', which suggests a desire for alliteration rather
than a potential truth of the role of carp in our rivers, lakes and wetlands. 'Victims' implies that
carp has done badly out of the situation, which couldn't be further from the truth. Didham
et al. (2005) contrasted the two roles instead as 'drivers vs passengers'. In our view, though,

'passengers' is a little passive, implying that species such as carp do not exploit the degraded environments but just go along for the ride. But perhaps we should all avoid too much anthropomorphising.

Although the 'Villains or victims?' pamphlet pointed out many of the ways carp can degrade a water body, it also emphasised that carp proliferated at least partly because of the way our rivers have been treated, and that native fish, in contrast, have struggled to compete. The carp has taken advantage of the degraded environment which it inhabits. The pamphlet had a foot firmly planted either side of the billabong. More than two decades on, we are still uncomfortably straddling the same body of water and the water remains depressingly murky, but less than it was.

Some might ask: 'Why do we need to know if it is carp or something else that is the problem? Surely getting rid of carp is a good thing?' Well, yes and no. Government coffers are not limitless, and we need to prioritise what funds are spent on. Notwithstanding, it is generally accepted that if an introduced, non-native species establishes itself, then spreads and becomes a successful invader, it will in all likelihood cause environmental problems (Gozlan *et al.* 2010). If the invasive species is an ecosystem engineer (one that directly or indirectly affects the availability of resources to other species and therefore modifies, maintains and/or creates habitats), as is the carp, then the effects are likely to be even more substantial and harmful. It is wise to be guided by the precautionary principle, which assumes that invasive species are a problem by definition, and so we should seek to prevent their introduction in the first place.

But, as we have just cautioned, we should be careful in ascribing *every* subsequent environmental change to the direct actions of an invading species. Whenever possible, we need to be guided by the evidence – and the evidence needs to be based on more than guilt by association. Working out whether the carp has been a 'driver' of environmental change or a 'passenger', or whether there is some 'back-seat driving' going on, depending on the environmental context, requires experiments in the field under realistic conditions, and preferably experiments where the species is removed and quantitative measures of recovery or non-recovery of native species are made (Didham *et al.* 2005). Furthermore, if environmental change is detected, for it to be considered harmful, those changes need to be more than superficial.

Fig. 8.1. Pamphlet 'Carp: Villains or victims?' Source: Murray–Darling Basin Commission and the National Carp Task Force, public domain.

In this chapter, we provide what evidence exists, first globally and then in Australia, on the impacts of carp in aquatic ecosystems. We touch briefly on the carp's impacts on the economy. Chapters 10–12 consider the measures used by some scientists, natural resource managers and the community to manage and control carp in Australia, while other groups have exploited the fish for commercial venture and recreational pastimes.

OVERVIEW OF THE IMPACTS OF CARP OVERSEAS

As we have said, the introduction of any non-native species will have an impact by its very existence. In many cases, the impact of the introduction may be relatively slight (e.g. tench in Australia), in other cases the impact may be profound (e.g. brown and rainbow trout). In broad terms, the impacts of introduced non-native fishes can be grouped into processes involving predation, habitat degradation, competition, hybridisation with native or other non-natives, and spread of disease (Gozlan *et al.* 2010). Whether an introduction is harmful – or likely to be in the future – is ultimately what we want to know, because then we can work out whether it is cost-effective to attempt to do something about it. And when we talk about 'harm', Gozlan *et al.* (2010) suggest the following:

> *The question is not about ecological changes as changes are inevitable when any species is introduced and established in a native ecosystem, but rather if these changes lead to a* measurable loss of diversity or change in ecosystem functioning. *Only in this case can fish species introduction be considered harmful (Gozlan* et al. *2010, p. 758).*

The carp has successfully established in approximately 90 countries of the 120-odd to which it has been introduced (Casal 2006). Although there have been many experimental studies investigating the effect of introductions of the species, both in the field and in the laboratory, in only a handful of cases have these been conducted at the scale of river systems and whole catchments – the scales with which most managers have to contend. The most comprehensive reviews – some of which include new data and/or models – of the effects of carp on aquatic systems are by Weber and Brown (2009), Matsuzaki *et al.* (2009), Badiou *et al.* (2011), Kulhanek *et al.* (2011), Vilizzi *et al.* (2015) and Fanson *et al.* (2024). These analysed the results of between 24 and 129 experimental studies. Combined, the studies encompassed more than 90 years of research (although mostly within the last 30 years), were globally distributed (although about one-third were from the US and many were from Australia and New Zealand) and included laboratory, field-based and 'natural' (unplanned serendipitous observations of natural events) experiments, with the vast majority in still-water habitats such as ponds and lakes. Vilizzi *et al.* (2015) also compared field enclosure experiments, in which fish were stocked in a confined space of some sort, with exclosure experiments in which carp were excluded from a confined space. It is apparent that the motivation and incidence of carp impact studies largely comes down to whether people see the fish as a problem rather than a resource (Vilizzi *et al.* 2015).

The meta-analyses (reviews that quantitatively analyse the results of many studies) of each review generally assessed whether there was a negative, neutral or positive effect of carp on some aspect of the environment. The most common components that were investigated were: 1) nutrients (usually nitrogen, phosphorus and sometimes ammonia); 2) turbidity (murkiness of the water) or suspended solids (material in the water column that may affect light penetration); 3) chlorophyll-a (photosynthetic pigment as a proxy for algal concentration) or phytoplankton (otherwise known as planktonic algae); 4) zooplankton (planktonic animals, usually comprising microcrustaceans and rotifers); 5) aquatic or semi-aquatic plants (collectively known as macrophytes); 6) macroinvertebrates (usually aquatic insects, e.g. mayflies, caddisflies, true flies, beetles and true bugs but also worms, large crustaceans and molluscs); 7) fishes; and 8) water birds.

It must be borne in mind, too, that the impacts of an invasive species like carp, while affecting one or several environmental features such as those above, can affect ecosystems at different scales (Britton 2023). The impacts can be on: 1) individuals (e.g. carp eat the eggs of a freshwater catfish, *Tandanus tandanus*); 2) populations (e.g. increased turbidity caused by carp feeding reduces or eliminates a species of macrophyte from a shallow lake); 3) communities (e.g. selective feeding by larval carp changes the community composition of zooplankton in a floodplain wetland); and ultimately 4) ecosystems (e.g. high densities of carp uproot all macrophytes in a floodplain wetland while feeding, and the wetland shifts from a macrophyte-dominated state to one where it is algal-dominated).

It is also important to consider that the effect of carp on environmental components can sometimes be direct, such as when a carp actually eats animals, thus reducing their numbers. Carp may also directly compete with other fish species for food (Mutethya and Yongo 2021). Often, however, the effect can be indirect, such as when a carp feeds on benthic (bottom-dwelling) macroinvertebrates. In the process, a carp may uproot macrophytes and resuspend material into the water column, increasing turbidity and reducing light penetration, which in turn can affect the growth of macrophytes or algae that were not directly disturbed.

Another consideration is whether the effects of carp: 1) are linear in nature (as density of carp increases, the environmental effects increase proportionally) (Fig. 8.2, line A); 2) are S-shaped (the effects of carp are non-linear and quickly increase at medium densities) (Fig. 8.2, line B); or 3) operate at thresholds (there is no effect until a certain density is reached, which can be low or high) (Fig. 8.2, lines C_1 or C_2) (Yokomizo *et al.* 2009). The issue of thresholds will come up later in this chapter, in the discussion of the impacts of carp.

Many of the quantitative reviews mentioned above, as well as others that were qualitative in nature (Hume *et al.* 1983; King 1995; Koehn *et al.* 2000; Koehn

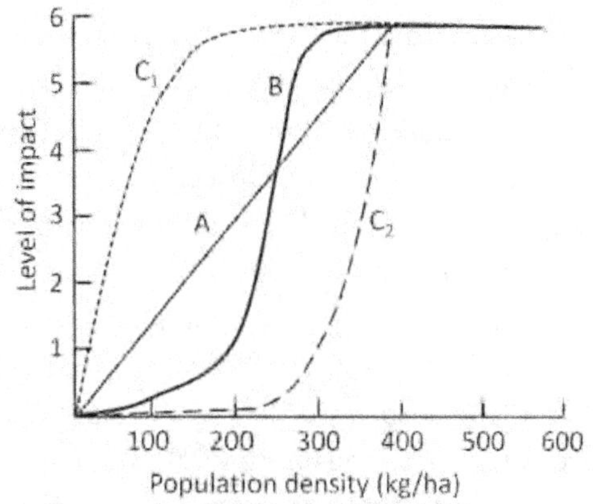

Fig. 8.2. Hypothetical relationships between density of an invasive species and the level of impact. (A) Linear. (B) S-shaped. (C_1) Low threshold. (C_2) High threshold. Source: Modified after Yokomizo *et al.* (2009).

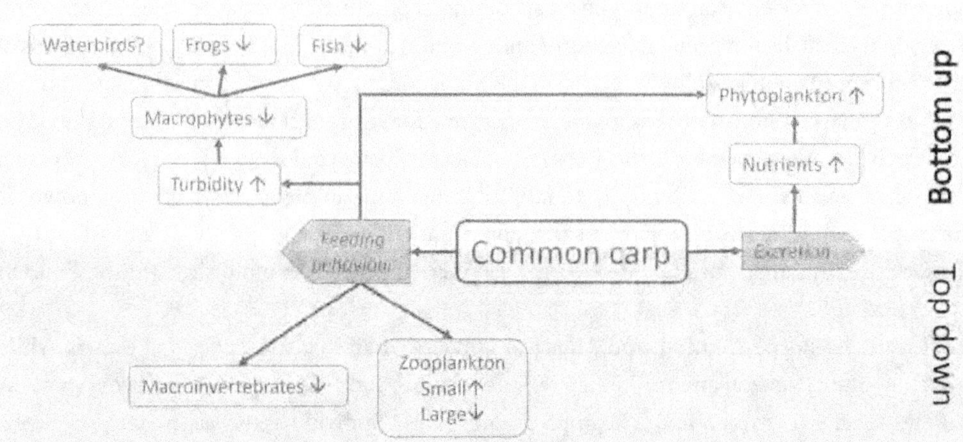

Fig. 8.3. The main top-down and bottom-up impacts of common carp on aquatic ecosystems. ↓ = decrease. ↑ = increase due to carp. ? = uncertain. Source: Modified after Hume *et al.* (1983), King (1995), Koehn *et al.* (2000), Weber and Brown (2009) and Vilizzi *et al.* (2015).

2004; Vilizzi *et al.* 2015), devised conceptual models which summarised the impacts of carp on ecosystems, and which we have adapted here (Fig. 8.3). The model has changed little since its inception. Broadly speaking, in aquatic food webs, fish can have top-down or bottom-up effects. The former is where the fish's feeding directly reduces the abundance of its prey, which has cascading effects down through lower levels. Bottom-up effects are where, as a byproduct of feeding, a fish may uproot macrophytes and resuspend sediments, creating turbid conditions that have flow-on effects at higher levels in food webs.

Weber and Brown (2009) concluded – and the other reviews cited above concurred – that carp has both top-down and bottom-up effects. They termed this a middle-out effect. Carp influences food webs directly from the top down as it feeds, reducing invertebrate abundance, diversity and assemblage composition. Carp also influences food webs from the bottom up through its feeding behaviour. It mobilises nutrients by resuspending sediments and uprooting macrophytes, and through its own excretion, and in the process increases turbidity. Remember, the overwhelming majority of studies were done in standing water bodies, such as ponds and shallow lakes. These are closed systems where the carp and the effects of the carp cannot escape; there is no flushing-out of sediment or turbid water or nutrients, which might occur in rivers. Notwithstanding the limited nature of such studies, experiments in the laboratory and in the field showed largely similar results, although these have not always been consistent between the two settings (Vilizzi *et al.* 2015). Context has been identified as critical (Kulhanek *et al.* 2011; Fanson *et al.* 2024). Overall, there is convincing evidence that carp: 1) increases levels of nitrogen and phosphorus; 2) decreases macrophyte cover; 3) increases turbidity and/ or suspended solids in the water column; and 4) reduces or alters the composition or diversity of benthic invertebrates and fish (Fig. 8.3). Evidence for the effects on phytoplankton and/or chlorophyll-a concentrations is moderate and on zooplankton it is inconsistent. Until relatively recently, there were few unequivocal studies on water birds. A study of the effects on water birds of an introduction, eradication and re-introduction of carp in two Spanish lakes, however, showed that carp can have a dramatic effect on herbivorous water birds through its effects on macrophytes, and on diving birds through competition for benthic invertebrates (Maceda-Veiga *et al.* 2017).

It is useful to unpack these results a little. In general, experiments indicated that, while excretions of fish that are benthic feeders can be positive, carp tends to overdo it, causing excess amounts of nutrients and algal blooms (Weber and Brown 2009). These effects tend to be more prominent with adult carp because of the way they feed, and in water bodies with soft sediments. Feeding by carp in soft sediments typically suspends material into the water column, increasing turbidity and reducing light penetration. This in turn reduces the growth of macrophytes, which may have already been disturbed or uprooted by foraging carp. This may cause problems for filtering invertebrates, such as zooplankton or visual feeding fishes. As carp can penetrate 12 cm into the sediment during feeding, it can disturb a lot of sediment. Most studies that investigated the impacts on turbidity showed an increase, and none showed the opposite (Weber and Brown 2009; Badiou *et al.* 2011; Vilizzi *et al.* 2015). Although Vilizzi *et al.*'s review showed only moderate support for increases in phytoplankton and chlorophyll-a, Weber and Brown's review showed that when carp was present, 80% and 71% of studies indicated increases in phytoplankton and chlorophyll-a, respectively. A combination of increased nutrients and turbidity can result in a change in the phytoplankton assemblage from green algae to blue-green algae (otherwise known as cyanobacteria), which can outcompete macrophytes and lead to a shift from a macrophyte-dominated water body to an algal-dominated one. Removal of carp can, in some cases, reverse this situation (Huser *et al.* 2022).

In contrast to the feeding of large carp, the species when small feeds mostly on zooplankton. It can affect the abundance and biomass both up and down, depending on the water body. Overall, however, small carp seem to affect zooplankton assemblages by selective feeding, with the result that the abundance of small zooplankton taxa increases at the expense of large species (Weber and Brown 2009). This may have flow-on effects on ecosystem functioning. Once carp switches to a benthic feeding existence at around 100 mm in length, evidence indicates a consistent reduction in the abundance and diversity of benthic invertebrates. Fish may also reduce numbers of shrimp and crayfish, possibly because of the effect carp has on macrophytes rather than through direct predation. Indeed, 96% of the studies reviewed by Weber and Brown showed a decline in abundance and diversity of macrophytes in the presence of carp. Macrophytes with good root systems, and those that live on harder substrates and are shade-tolerant, fare better than macrophytes with the opposite characteristics.

Some studies have shown that the properties of water bodies are affected once the density of carp reaches certain thresholds (lines C_1 and C_2 in Fig. 8.2), although this is very much context-dependent (Weber and Brown 2009; Kulhanek *et al.* 2011; Britton 2023). Others have concluded that there are linear relationships along the full range of carp biomass with many of the variables (line A in Fig. 8.2), except – depending on the meta-analysis – suspended solid concentrations, turbidity, macroinvertebrates and macrophytes (Kulhanek *et al.* 2011; Fanson *et al.* 2024). Studies in closed systems that mention thresholds of significance, indicate that carp densities of 250–450 kg/ha need to be reached for effects on ecosystems to be prominent (Weber and Brown 2009). However, effects at densities as low as 5 kg/ha have been detected (Richardson *et al.* 1990), while in other cases densities as high as 750 kg/ha were needed to produce an effect (Lougheed *et al.* 1998). The meta-analysis of Fanson *et al.* (2024) indicated that by the time carp biomass in a water body had reached 250 kg/ha, about 50% of the macrophytes and macroinvertebrates would have been lost. The size of fish, type of substrate, size of the water body, whether it was closed or open and whether the experiment was carried out in the laboratory or field all influenced critical thresholds in experimental studies.

THE AUSTRALIAN EXPERIENCE

Australia has an enviable history of investigating the effects of carp on the aquatic environment since the BEH. An article and Honours project focusing on Gippsland in the late 1960s and early 1970s kicked off assessments of the effects of the Boolarra carp (Roberts 1969; Malcolm 1971). But it was the appointment of Fred Reynolds in South Australia in the mid-1970s that initiated the first substantial program of work that aimed to investigate the biology and habits of carp in Australia. Reynolds' research, especially into the movement of carp (Chapter 6), while suffering from the lack of technology which we now take for granted, provided insights that were broadly confirmed by, and paved the way for, future research. Reynolds' work was just the beginning. A wave of studies was to follow – indeed, work is ongoing.

Because the number of carp swimming around doing their thing will have a bearing on how widespread the impact is, we first look at estimates of fish numbers. Ivor Stuart and colleagues explored data spanning 1994–2018, based on collections of more than half a million carp from 4,831 sites and 153 studies wherever in Australia carp have been found (Stuart *et al.* 2021). They came up with an average estimate of 199.2 million carp weighing 215,456 tonnes when flows were average, and a lot higher if conditions were wetter! While that might seem like a lot of carp – and it is – it is probably even more sobering to realise that carp likely occupy 92% of aquatic environments in the Murray–Darling Basin and the south-east coastal divisions of Australia. Stuart *et al.* estimated that 56% of wetlands and 70% of rivers likely have densities of carp exceeding 80–100 kg/ha. Another study from Hayden Schilling and colleagues, also making use of a large dataset from 1994–2023, estimated that carp in New South Wales sites of the Murray–Darling Basin make up approximately 57% of the current fish biomass and that this proportion has not changed much over the last 30 years (Schilling *et al.* 2024).

Timeline of studies

Broadly speaking, there are five approaches to assessing the impact of a species like carp on the aquatic environment: 1) anecdotal observations; 2) scientific surveys; 3) controlled experiments that involve enclosures or exclosures; 4) removal and recovery experiments; and 5) modelling studies (Fig. 8.4). The first approach, while usually the most common source of 'evidence', is ultimately not useful scientifically because results are unverifiable unless meticulous before-and-after records are kept. The second approach is by definition more scientific, but both it and the first approach suffer from an inability to tease out environmental factors other than the

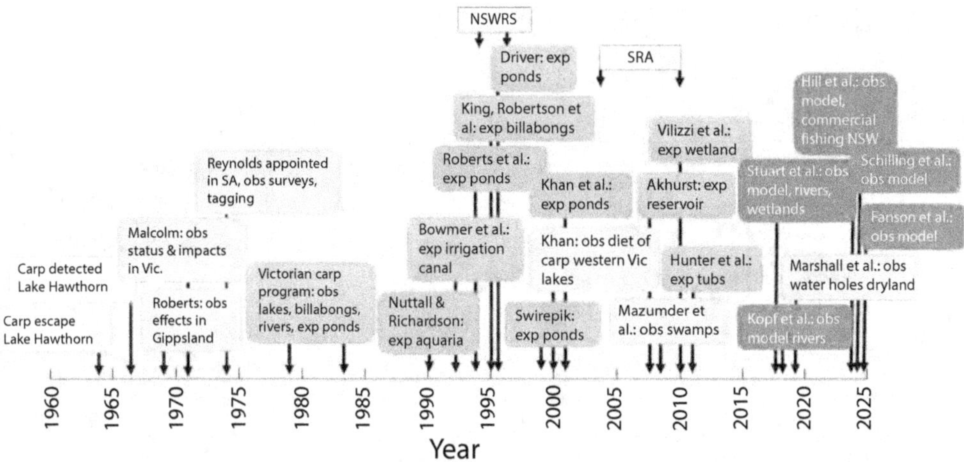

Fig. 8.4. Timeline of selected studies of the impact of carp in Australia. Dates are approximately when studies were conducted unless no date was given or the study used data over many years, in which case the date of publication is indicated. If work originated from a thesis, the thesis date is indicated if possible. Exp = experimental study; obs = observational study; obs model = modelling study based on observational data (commercial fishers' observations started early and continued for many years); NSWRS = NSW Rivers Survey; SRA = Sustainable Rivers Audit.

presence of carp (Brown 1996). The third approach can control for other environmental factors but it often lacks realism unless carried out in natural waters, so the opportunity to scale-up to real-world situations may be limited. The fourth is one of the best approaches, because, if measurements are made *before* removal, these can be compared with *after* removal of carp. If the experiments are carried out in the field, the results can probably be transferred to other situations with a degree of confidence. The problem with many such experiments, as we have seen above, is that they are difficult to carry out successfully and are usually conducted in closed systems such as ponds and shallow lakes, which may have questionable relevance to open, flowing-water environments. The last approach, modelling, can be powerful, especially if it includes data over long periods or large geographic areas. In many cases it is the best we can do because of the spatial scale of the problem. Modelling typically takes real data, preferably from a range of sites that have a gradient of carp densities and (possibly) unrelated environmental variation, and estimates the relative impact of not just carp, but other potential influencing factors.

Below, we describe some of the most useful carp studies of the last 40 years or so, that have directly investigated the impacts of carp in Australia. We don't dismiss the importance of anecdotal accounts, but they have rarely been collected systematically (but see Roberts and Sainty 1996) and they cannot be checked for veracity, so we mostly avoid them for those reasons. The reports of professional fishers have been absolutely essential at times in alerting government agencies to issues that they should be concerned about. Keith Bell of K & C Fisheries Pty Ltd has been an especially important source of information and advice for decades (see Chapter 10). There are some compelling written anecdotes of the effect of carp on macrophytes (Sinclair 2001) and on freshwater catfish (Frawley *et al.* 2012).

Fig. 8.5. A montage of Victorian Carp Program photos, 1980. (A) Carp program team, left to right: Greg Smith, Andrea Fletcher/Brumley, Dick Brumley, Bruce Allsop and Doug Hume. (B) Andrea Fletcher/Brumley driving a boat on the Goulburn River. (C) Seining for carp in Lake Cooper, Dick Brumley in waders. (D) Doug Hume and Jim Pribble measure a carp for an article in *The Australasian Post*. (E) Floy-tagging a carp. (F) Wanganui Ponds, Shepparton. Source: Doug Hume.

Victorian Carp Program

The next major enterprise after Reynolds' appointment in South Australia was the Victorian Carp Program, which ran from 1979 to 1982 (Hume *et al.* 1983) (Fig. 8.5). It got off the ground after a 1976 proposal to the Victorian Government by Jim Wharton of the Victorian Fisheries and Wildlife Division, following pressure from angling and hunting groups (Peter Rogan, pers. comm.). American fisheries scientist Jim Pribble was appointed as Program Scientist; further staff included Andrea Fletcher/Brumley, Sandy Morison, Alisdair Brown, Dick Brumley, Doug Hume and several others. Pribble appointed Hume because few people at the Arthur Rylah Institute in those days had a background in fisheries, and Hume (a Sydney-born Australian) had spent several years in the US studying that subject (Doug Hume, pers. comm.). Hume was instrumental not only in overseeing the field work but in standardising the reporting of results and trying to make the data freely available. The latter was the norm in the US but was yet to be adopted as routine (and indeed, was viewed with scepticism) in Victoria. The Carp Program initially operated out of an old auto-mechanics establishment in Shepparton, central Victoria, before the Kaiela Research Centre was set up. We are very fortunate to have the 10 detailed reports produced during those three years, which show intense activity and an enviable record of accomplishments in a comparatively short time (see Hume *et al.* 1983). The report topics included, in order: 1) a proposal to assess the impacts of carp in Victoria; 2) a bibliography of carp studies; 3) a review of the impacts of carp on fish and invertebrates; 4) a review of the impacts of carp on macrophytes and water birds; 5) a review of the biology and behaviour of carp; 6) genetics and the management of carp in Victoria; 7) annual reports for 1979–1980, 1980–1981 and 1981–1982; and 8) a final report summarising the program and its results.

In addition to reviews of the impacts of carp, bibliographies of studies and assessments of the potential of genetics, the Victorian Carp Program conducted experimental and observational studies on the ecological impacts in Victorian ecosystems, both in mesocosms and in the field, and trialled ways to remove carp. For our purposes, the most important projects were two that looked at the impacts of carp. The first was a set of experiments at Wanganui Ponds at Shepparton, Victoria. Four 1 ha ponds were divided into quarters and stocked with carp, including controls with no carp (Hume *et al.* 1983). There were problems with preventing movement of carp between sections of each pond, and evaporation and refilling confounded results somewhat, as did excessive growth of water couch (*Paspalum distichum*) and predation by water birds. Overall, results showed that conductivity (the saltiness of the water), temperature, pH and dissolved oxygen were more affected by water levels and seasonal conditions than by the presence of carp. There was limited evidence that zooplankton abundance declined for some species, but again, season had a large effect. Results for the impacts of carp on native fish could not be assessed. The results for macrophytes were equivocal, but like those of most studies, indicated that shallow-rooted, fragile, submerged species were affected more than were more robust and emergent species.

The Victorian Carp Program's second study was mostly observational. It investigated the effect of carp in a number of billabongs near Shepparton and the Broken River (a tributary of the Goulburn River, northern Victoria). The researchers stocked Waugh Road Billabong with extra carp to investigate the effects. Results indicated that turbidity was related to hydrology and not to carp, and that fragile macrophyte species such as *Potamogeton* spp. and *Chara* spp. were potentially eradicated from systems in which carp were found, but hydrology and water level had important influences on these too. There was no evidence that water birds were affected by carp. Although the results were patchy in places and there were some procedural issues with experiments, the investigations of the Victorian Carp Program overall were groundbreaking and provided the benchmark for later studies. The focus on carp and the sheer volume of results is unparalleled.

Studies of the impacts of carp in Australia

Following the Victorian Carp Program, a series of experiments and observational studies started in the early 1990s and continued for the next few decades (Fig. 8.4). Notable among the studies were:

- an investigation of the potential use of irrigation canals as wetlands and how carp would affect this (Bowmer *et al.* 1994);
- a series of experiments in ponds at Griffith which looked at the effects of low and high densities of carp on a range of water quality variables, nutrients, phytoplankton and macrophytes (Roberts *et al.* 1995);
- an enclosure/exclosure experiment in billabongs on the Murrumbidgee floodplain, again looking at the effects of high and low densities of carp on environmental aspects under more natural conditions (King *et al.* 1997; Robertson *et al.* 1997);

- a study in outdoor ponds on the effects of carp on macrophytes (Swirepik 1999);
- an observational study of the overlap of diets of carp of different ages and sizes with those of native fish, and an experiment investigating the effects of high densities of carp on nutrients, phytoplankton and zooplankton, both in western Victorian lakes (Khan 2003; Khan *et al.* 2003);
- an enclosure study in a billabong that looked at the effects of carp density and size on water quality, nutrients and phytoplankton (Driver *et al.* 2005a);
- experiments in tubs that investigated the predatory effects on tadpoles of the Boorolong frog, *Litoria booroolongensis* (Hunter *et al.* 2011);
- a comparison of the diet of carp and native carp gudgeons in the Lowbidgee floodplain (Mazumder *et al.* 2012);
- a northern New South Wales experiment in enclosures in a reservoir that looked at the effects of carp, gambusia and native fishes on water quality, nutrients, algae, macrophytes and zooplankton (Akhurst *et al.* 2012); and
- a series of exclosure experiments over two years, examining the effects of carp in lower Murray floodplain wetlands, on the usual suspects, as well as on benthic macroinvertebrates and fish (Vilizzi *et al.* 2014).

The studies were of closed systems, some artificial and some natural; all broadly showed the same results and mirrored those found overseas (Weber and Brown 2009). The similarities were striking. Most studies investigated the effects of carp on water quality variables such as turbidity and pH, many on nutrients, phytoplankton, macrophytes and zooplankton, only a very few on fish and none on water birds. No experiments looked at recovery of ecosystems after the removal of carp. When removal experiments were done, these were at very small scales, such as transferring carp from one side of bisected billabong to the other (King *et al.* 1997). We will necessarily gloss over detail, as we do not have the space here to include all the subtleties of each study. We stress that, although most studies agreed on the direction of change, some failed to detect a change. Importantly, only one to our knowledge showed changes in opposite directions, and it was most likely due to reasons other than carp.

Overall, in the presence of carp under experimental conditions, sediments are usually suspended if substrates are soft, and turbidity and nutrients (usually nitrogen and phosphorus) also increase. Larger fish suspend more sediment by their feeding action than do smaller fish (Driver *et al.* 2005a). As a result, phytoplankton abundance (or a proxy, concentrations of chlorophyll-a) normally increase, but not always (e.g. Roberts *et al.* 1995). Shallow-rooted submerged macrophytes tend to decline in cover and abundance, whereas more robust emergent macrophytes do not (Roberts *et al.* 1995), and macrophytes may be especially vulnerable after water bodies have dried and/or when macrophytes are recruiting (Swirepik 1999). Zooplankton abundance may or may not decrease in the presence of carp, but it is more likely that zooplankton composition will change through selective feeding (Khan 2003). The diets of carp are majorly affected by water level in wetlands (Mazumder *et al.* 2012). Carp will eat tadpoles when available and refuges do not offer protection from that predation (Hunter *et al.* 2011). Carp rarely eat fish eggs or fish (Khan 2003). When the effects of carp on benthic macroinvertebrates were investigated, there were mixed results. Carp certainly eats them (Hume *et al.* 1983; Khan 2003), but then so do many species of native fishes. Vilizzi *et al.* (2014) found a weak negative relationship of carp biomass with the number of taxa (but not the density or diversity) of benthic macroinvertebrates. This could be a direct result through feeding, or indirect through the effect of decreased water transparency or degradation of the macrophyte habitat, according to the authors.

Carp, native fish and big data

As we have said, there have been very few experiments involving native fish and none to speak of on birds, which is not all that surprising because of the time and space needed for such investigations. The exception for fish is the study by Vilizzi *et al.* (2014). However, this was a rare case when results were the opposite of what would be expected: native fish biomass (mostly bony herring, *Nematalosa erebi*) increased as carp biomass increased, although there was only a weak association and it was more likely the direct result of flooding and not because of the beneficial effects of the presence of carp. This is a good example of how correlation does not mean causation: both carp and bony herring were responding positively to an unrelated factor, flow. Although not an experiment, a noteworthy study by Jon Marshall and colleagues investigated, among other things, native fish biomass and abundance in seven dryland river catchments, some with and some without carp (Marshall *et al.* 2019). Although geography was most influential in predicting the abundance and biomass of native fish, carp apparently was associated with a decrease in native fish biomass, likely because carp dominates food sources and is a long-lived sink for energy. Marshall *et al.* also noted the loss of a species of aquatic snail (*Notopala sublineata*) and a decrease in macrophyte occurrence in the presence of carp, although macrophytes in these dryland rivers are rare naturally.

All carp removal and recovery experiments, as far as we are aware, have been conducted overseas (e.g. Bajer and Sorensen 2015), although methods for removal have been compared in Australia (Norris *et al.* 2014) and in the case of Tasmania have been been successful (see Chapter 9). For large-scale assessments of the effect of carp on mobile and long-lived animals like fish, data modelling come into its own.

Since carp spread throughout the Murray–Darling Basin and elsewhere, virtually every fish survey has caught carp. The New South Wales Rivers Survey of 1994–1996 (Harris and Gehrke 1997) and the Sustainable Rivers Audit of 2004–2007 and 2008–2010 (Davies *et al.* 2010) collected data on carp and indeed all freshwater fish. The fisheries organisations of most jurisdictions in the Murray–Darling Basin have routinely collected fish, including carp, over several decades. The results from such studies provide a wonderful opportunity to use the data on carp, native fish and any other measured variables to assess associations of carp with environmental conditions more generally.

Keller Kopf and colleagues (with PH as a co-author) conducted a modelling exercise, based on Sustainable Rivers Audit data from 251 sites in 17 catchments (Kopf *et al.* 2019b). Some catchments had no carp while others had very large numbers of carp, some had no flow alteration while others were heavily modified. Using ecological theory about the way that body size and biomass of animals are related in ecosystems (usually small-bodied organisms make up the largest biomass at the bottom of a food pyramid) and how energy moves up the pyramid, we estimated that carp and river regulation have contributed roughly equally to a 58% decline in native fish biomass. Our study was the first attempt to put an actual number on the decline. We hypothesised that the carp had affected the pyramid because of its large body and where it feeds – at the bottom of the food web. We also estimated that reduction of carp to 30% of the fish assemblage would have the potential to more than double the current native fish biomass. Our study suggested strongly that river regulation *and* carp are both problems that need to be addressed if we hope to restore our rivers (Kopf *et al.* 2019b).

An extensive dataset from 1994–2018 (state agency records from Victoria, Queensland, Australian Capital Territory, New South Wales and South Australia, and presence/absence data from Western Australia and Tasmania) was used in an assessment of the effects of carp on a range of ecological components of rivers (Fanson *et al.* 2024). The authors included rivers and wetlands in their analysis, developed correction factors for sample detection efficiency and drew on carp removal experiments to estimate the effects of carp and model error. Fanson and colleagues estimated that, because of carp, macrophytes have declined in Australia by 41%, generally more in wetlands and lakes than in rivers. In no case did macrophyte coverage increase. They also confirmed that suspended solids, nutrients, turbidity and phytoplankton had all increased because of carp. The effect on macroinvetebrates was equivocal and fish were unfortunately unable to be tested.

Economic impacts of carp in Australia

As a bit of an aside, although very relevant to this chapter on impacts of carp in Australia, is the question of economic impact. Rarely has this been calculated. In 2004, the Cooperative Research Centre for Pest Animal Control (CRCPAC) commissioned Ross McLeod of eSYS Development Pty Ltd to estimate the economic costs of carp in Australia. Included were control costs, production losses, research and management costs and environmental costs through impacts on biodiversity (McLeod 2004). The overall economic impact of carp as of 2004 was estimated to be A$15.8 million annually. This comprised A$2 million spent in management, A$2 million in research and an estimated A$11.8 million in losses due to environmental degradation, mostly through increased turbidity, loss of aquatic macrophytes and damage to irrigation canals. But this needs to be offset with the economic benefits of the sale and processing of carp for other purposes, such as fisheries for various purposes, and fertiliser (see Chapter 10). Two other studies that made comprehensive assessments were one by the Gippsland Lakes and Catchment Action Group in 1996 and another by the National Carp Control Plan in 2019. The first estimated that in the mid-1990s there was an annual cost of A$35 million to commercial and recreational fishing, tourism and commerce in the Gippsland Lakes community alone (Koehn *et al.* 2000). The second study was unable to define a 'non-market' cost for Australian households, based on how much people were willing to pay to rehabilitate the environment once carp numbers were reduced (Hardaker *et al.* 2019). The amount that people were willing to pay depended very much on projections of the likelihood and extent of environmental recovery if carp numbers declined because of the cyprinid herpesvirus CyHV-3 (see Chapter 10). McLeod did not attempt to estimate the social cost of carp in Australia.

UNFILLED KNOWLEDGE GAPS AND CARP AS 'DRIVERS' OR 'PASSENGERS' REVISITED

Despite the flurry of activity over the last 40 years or so, there remain many gaps in our knowledge of the impacts of carp in Australia. This is not surprising. After all, we are still in the early days of the carp invasion. Give it a couple of centuries or so, then we will revisit what we know. In our view, current gaps in our knowledge are:

1. the impacts of carp in open, flowing-water systems. The vast majority of studies have been conducted in shallow lakes, billabongs and other floodplain wetlands or under controlled laboratory or mesocosm conditions. With the advent of long-term big datasets and the modelling this allows, gaps are starting to be filled and the future is looking brighter;
2. the impacts of carp on aquatic food webs. Most studies have investigated different components of the aquatic environment, but not combined these in a systematic way that gives meaningful insights into how carp affects food webs as a whole. Perhaps only the study by Keller Kopf and colleagues (2019b) has attempted to integrate the biomass of carp into how 'natural' food webs are thought to operate. More needs to be done;
3. the wildlife that eat carp. Few studies have considered the role of carp as food for wildlife, whether that be other fish, birds, water rats or terrestrial mammals. The role may be positive or negative. Katie's PhD thesis (Doyle 2012) and a study by Heather McGinness and colleagues (McGinness et al. 2019) are rare examples of this research;
4. the impacts of carp on terrestrial ecosystems. To our knowledge, there have been no studies of the flow-on effects of the occurrence or density of carp on the terrestrial environment in Australia, yet because of the carp's readiness to move into temporary floodplain wetlands, its effects especially on aquatic–terrestrial transition zones are likely to be profound;
5. carp's effects relative to other fish species. There has been a dearth of studies or even questions related to the relative impact of carp versus native fish species. For example, if enclosure experiments on bony herring were conducted like they have been on carp, would that species be shown to have similar impacts on turbidity, nutrients and phytoplankton as does carp;
6. the impact of carp on native fish more generally. As we have pointed out, there have been few studies on the effects on native fish, and the results of those that have been done, have typically been equivocal.

Interestingly, this last 'gap' may be where anecdotal evidence may come in handy. A series of interviews – *Talking Fish* – with people living along the rivers of the Murray–Darling Basin, carried out by Jodi Frawley, Scott Nichols, Heather Goodall and Liz Baker, documented stories of their connections with and ideas about fish (Frawley *et al.* 2012). There were lots of comments about the condition of rivers before and after carp arrived on the scene. Again, there is the problem that other river management impacts may have happened in concert with the arrival of carp. But the people interviewed repeated many times that since the spread of carp, freshwater catfish numbers have declined, presumably because carp eats catfish eggs or in other ways disrupts the nests. Ironically, probably the best evidence for carp's impacts on freshwater fish in Australia is its presumed role in the demise of another non-native species, tench (Hill *et al.* 2024). Not an anecdotal study, but a big data one.

Finally, virtually no Australian studies that have looked at the impacts of carp have attempted to compare the relative impacts of carp with other environmental impacts, such as flow alteration, habitat destruction, cold-water pollution and the rest of the many problems that beset our rivers, floodplains and lakes. Therefore, while we can unequivocally say that carp affects the ecosystems in which it is found, and that the greater the density of carp, generally the greater the impact, there is still a degree of uncertainty as to how much degradation is due to carp and how much is due to other factors. Is carp a driver, a passenger, both, or something in between? In our view, carp is best described as a back-seat driver: it takes advantage of environmental conditions, either natural or more commonly anthropogenically, modified. It proliferates and then through its behaviour creates conditions in which it can increase further. In the process, the altered conditions we humans initiated and carp exacerbated add insult to injury for many native fish species (and other components of the environment). This has been the pattern elsewhere, most notably in western Europe, and there is little reason to think that the Australian experience is unique. The perfect storm – the BEH and subsequent explosion of carp in the Murray–Darling Basin coinciding with a rapid rise in storage capacity and flow alteration – may be a coincidence, but we doubt it. The one spark of doubt is that the Boolarra carp has also proliferated in Gippsland. And so, the murkiness that envelops our understanding of carp in Australia persists.

*****

More research and similar assessments need to be done to tease out the relative impacts of carp versus other sources of environmental degradation. We are not saying it is easy. It clearly isn't, because otherwise such studies would have been done by now. But with the large amounts of data that have been collected over the last few decades, the prospect for such analyses is promising. However, there is still some haziness about where we would get the best bang for our buck in environmental management. Is attempting to dramatically knock down numbers of carp in Australia worth the effort environmentally, economically and socially? Ultimately, the community will decide. We revisit this in the final chapter of the book, where we combine all of what we have learnt about the history of carp introductions, biology, movement and spread in various parts of the world, including in Australia. For the moment, in the following chapter, we take a short trip away from mainland Australia and tell a rare good-news story of the rise and fall of carp in Tasmania.

9

Carp in Tasmania: A success story

'I hope the Commission can successfully solve the problem'. Don't we all, Graeme!
(Inland Fisheries Commission 1995).

CARP ORIGINALLY MADE ITS WAY TO MAINLAND AUSTRALIA VIA TASMANIA and, ironically, carp has made the reverse trip several times ... each time with the help of misguided people. It was reintroduced to the island between 1962 and 1964 (Inland Fisheries Commission 1975) and possibly again sometime prior to 1980, although the 52 fish found at that time could have been descendants of fish from the 1960s introductions (Inland Fisheries Commission 1980). The most recent introduction was discovered in 1995 and it is most likely that fish had been introduced some years before. In the 1960s and 1980 cases, the fish were discovered by officers of the Tasmanian Inland Fisheries Commission, and were quickly dealt with by poisoning. The 1960s fish we know were obtained from the Boolarra Fish Farms. The most recent 1995 introduction of carp was to two Tasmanian central highland lakes, lakes Crescent and Sorell, which catalysed frantic efforts to control and eradicate the species. These efforts took almost 30 years, but thankfully they have paid off. This was one of the few successful eradications of carp anywhere in the world.

So, the story of carp in Australia likely started in Tasmania and in a way has an ending there too. At least in a relatively small way, and at least for now. It shows what can be done if carp is contained, adequate funding is provided to understand the habits, population structure, breeding and movement of the species, and importantly, there is a concerted and adaptive approach to eradication. While mistakes were undoubtedly made during the process of containment, fishing down and eradication, that is how science and the management of invasive species works: learn along the way, improve what is done and get a better result over time.

This chapter charts that success story. We take you from the first finding by an angler of a half-eaten carcase at Lake Crescent, to the initial frantic efforts to contain the spread of the species to the sister lakes of Crescent and Sorell and to dealing with public, conservation and angling concerns. We describe the multiple innovative ways of finding and removing juvenile and adult carp that were developed, which also limited breeding and recruitment so that new generations did not come through. Yes, the situation is a lot less complex than that of mainland Australia and the Murray–Darling Basin in particular, but there are lessons to be learnt from the Tasmanian eradication. In what follows, we draw on many publications by the Tasmanian Inland Fisheries Commission (IFC), from March 2000, Inland Fisheries Service (IFS), particularly its annual reports as well as papers and reports published by IFS scientists, especially John Diggle, Paul Donkers, Jawahar Patil, Chris Wisniewski and Jonah Yick (Donkers 2004; Inland Fisheries Service 2004; Diggle *et al.* 2012; Donkers *et al.* 2012; Taylor *et al.* 2012; Wisniewski *et al.* 2015; Yick *et al.* 2021). For context of the times, we were fortunate also to be able to talk first-hand with many of those intimately involved: John Diggle, Wayne Fulton, Andrew Sanger, Chris Wisniewski and Jonah Yick. Our account mostly follows the on-ground work, and mentions only in passing engagement with the general public. But of course this is absolutely essential if eradication efforts are to be successful. From 1996–2004 alone, the Tasmanian Carp Management Program team gave over 60 public awareness presentations and there were more than 200 newspaper, magazine and TV articles that appeared in Tasmania and elsewhere (Inland Fisheries Service 2004). Carp in Tasmania was big news.

A CARP IS FOUND

On 28 January 1995, while fly fishing with his wife Maureen from a boat in Lake Crescent in the central highlands of Tasmania (Fig. 9.1), Graeme Porter spied a bird eating something on the bank and went to investigate:

> *... we noticed a magnificent sea eagle sitting on the rocky edge of the lake. As we drifted closer the eagle flew off. I was curious as to what the eagle had been eating and, upon investigation, I found an unusual fish with part of its head eaten. I wrapped the fish in a wet towel and returned to camp. We then located a Commission inspector at Dago Point, gave him the fish, and the rest is history (Graeme Porter, fisherman, Inland Fisheries Commission 1995, p. 7).*

Unbeknown to Graeme, he had found the first of many common carp that had been living and breeding in Lake Crescent and nearby Lake Sorell for probably four or five years (Inland Fisheries Commission 1995). The fish was positively identified as common carp, *Cyprinus carpio*, two days later. Within another couple of days IFC officers were at Lake Crescent and found more carp, some >1 kg in weight. On 18 February, the difficult decision was made to close Lake Crescent to the public (Inland Fisheries Service 2004). On 9 March, after carp had been found in Lake Sorell a few days earlier, it too was closed. Fish surveys in the lakes started immediately upon discovery of carp. Surveys downstream in the River Clyde, the outlet from Lake Crescent, began shortly after.

The IFC moved fast, knowing there was no time to lose. Although there was the inevitable feeling of doom and despair among IFC staff in the early days, then Commissioner Wayne Fulton was stubbornly resolute and pushed ahead with the drive and single-mindedness for which he was well known (Andrew Sanger, pers. comm.). A Carp Management Program was quickly formed within the IFC. To make the job of surveying the lakes more efficient, an electrofishing boat was borrowed from New South Wales Fisheries. This arrived on 20 February, and another borrowed from Victorian Fisheries arrived on 4 March. The Carp Management Program bought its own boat from the US and this arrived in early May. In the meantime, screens were installed in the outlet to Lake Crescent in an attempt to prevent the escape of carp downstream. The frenzy of activity was understandable, given what was known about the spread and impacts of carp on the mainland. But it was trebly worrying because of the recreational, commercial and conservation significance of the pair of lakes.

To put the situation in context, we need to know a little about the two lakes at the centre of this story. They are large and shallow freshwater lakes in the south-eastern part of the Central Plateau of Tasmania. The River Clyde flows approximately south-west from Lake Crescent, eventually reaching Meadowbank Lake and the River Derwent. Lakes Crescent (area 23 km²,

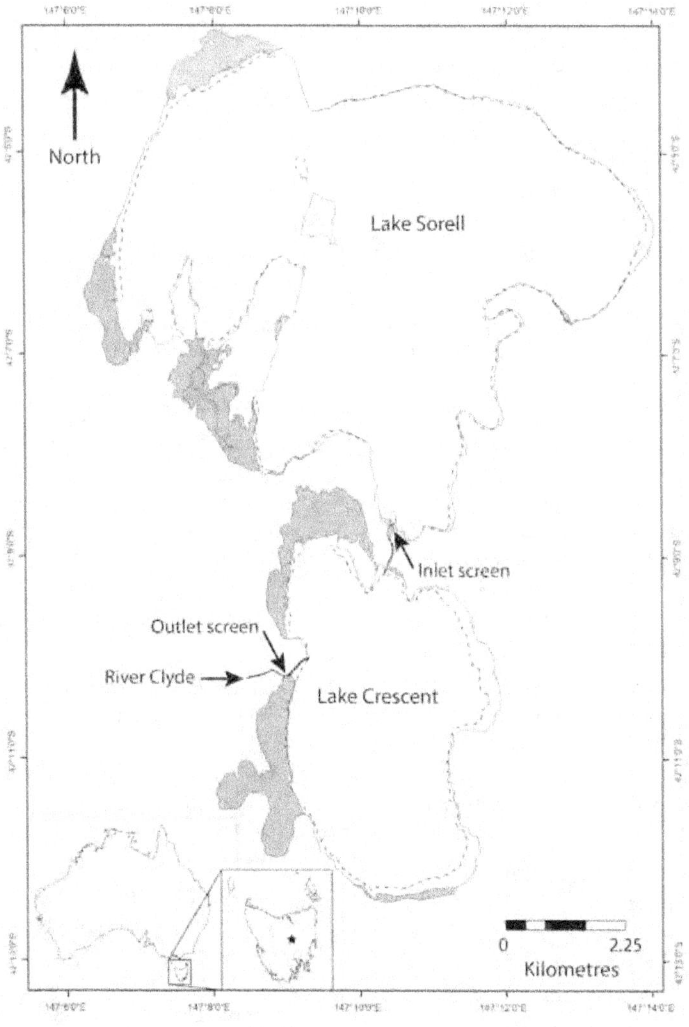

Fig. 9.1. Map of lakes Crescent and Sorell and the River Clyde. Source: Jonah Yick, Tasmanian Inland Fisheries Service.

maximum depth 3.8 m) and Sorell (area 54 km², maximum depth 4.4 m) were then – and now – home to the endemic and endangered golden galaxias (*Galaxias auratus*) and the endemic snail *Austropyrgus* sp., included Ramsar-listed wetlands at Interlaken, acted as important habitat for local and migratory water birds, supported many native endemic aquatic and terrestrial species, contributed to the lucrative Tasmanian brown (*Salmo trutta*) and rainbow trout (*Oncorhynchus mykiss*) recreational fisheries and supported a commercial short-finned eel (*Anguilla australis*) fishery (Inland Fisheries Service 2004; Yick *et al.* 2021). Water from the lakes was used for domestic and irrigation supply. In short, many groups were invested in the lakes and wanted to avoid their degradation by this new invader.

THE PLAN

Shortly after the discovery of carp in Lake Crescent, a Carp Task Force was set up (later known as the Carp Working Group), comprising representatives from the IFC, Rivers and Water Supply Commission, Bothwell Council, Clyde Water Trust, Freshwater Anglers Council of Tasmania and some independent parties (Inland Fisheries Commission 1995). It developed the initial and later strategies for dealing with the unfolding issue and with communicating information more broadly. The overall objective was 'To eradicate carp from Tasmanian waters and, in the meantime, to minimise the impact of carp on Tasmania from economic, recreational and ecological points of view' (Inland Fisheries Service 2004, p. 11). The plan included a number of

Table 9.1. Original and expanded objectives of the Carp Task Force in relation to carp management and eradication in lakes Sorell and Crescent (Inland Fisheries Service 2004, pp. 11–12).

Initial objective	Expanded objective
I) Contain carp in lakes Sorell and Crescent catchment	1) Contain carp in lakes Sorell and Crescent
II) Develop a water management plan that provides for and protects the water supplies for Bothwell, Hamilton and irrigators and achieves I above along with satisfying III and IV below.	2) Refine the water management plan to better provide for and protect the water supplies of Bothwell, Hamilton and irrigators
III) Reduce the existing carp population	3) Reduce the existing carp populations
IV) Eradication of carp	4) Eradication of carp
V) Prevention of reintroduction of carp to cleared waters from both interstate and intrastate	5) Prevent the reintroduction to cleared waters from both interstate and intrastate
VI) Undertake a communications strategy to minimise damage to tourism, and to support above strategies I to V	6) Undertake a communications and education strategy to minimise damage to tourism and increase awareness in fishing and general communities
	7) Improve the capacity of the containment screens
	8) Protect native flora and fauna threatened by carp or carp management
	9) Gain an understanding of factors controlling the success of carp in lakes Sorell and Crescent
	10) Develop guidelines for recreational and commercial access to lakes Sorell and Crescent

ambitious objectives (Table 9.1), the first of which was to keep carp confined to the two lakes, as this would buy time to develop other strategies for how to deal with the problem. The second objective was to maintain water supply to downstream users. Subsequent objectives were to bring down carp numbers in the lake and eventually eradicate them entirely, while ensuring that carp was not reintroduced from within and outside of Tasmania.

Considering the disastrous consequences if carp were to escape the lakes and disperse down the River Clyde to the River Derwent and beyond, containing the existing fish to the two lakes was an obvious priority. The first part of the solution was to close Lake Crescent to the public, because of the fear that people might deliberately or inadvertently spread carp beyond the lakes (Inland Fisheries Service 2004). The second part was to stop carp moving through Kermodes Cut, a channel between the two lakes. A vertical grate was used at first, replaced by a 5 mm screen in 2001. Finally, carp had to be prevented from moving out of Lake Crescent through the lake outlet into the River Clyde. Fortunately, the outlet is relatively small and so 1.1 mm screens were installed to prevent the movement of carp of all sizes, except perhaps carp larvae, whose head width might have allowed them to penetrate the small aperture (Fig. 9.2). There were teething problems with the screens and for the first year or so, the amount of water able to be released was limited and Lake Crescent filled to levels that risked providing good conditions for carp spawning. Later, 5 mm screens replaced the 1.1 mm ones during months when rainfall was high, temperatures were low and small carp were absent. The eventual configuration included double screens of coarse followed by fine mesh.

Fig. 9.2. The Lake Crescent outlet structure showing the coarse external screens. Source: Jonah Yick, Tasmanian Inland Fisheries Service.

But because blockage was a persistent problem, daily maintenance and cleaning of the screens were necessary.

Managing the levels of the lakes was an important strategy to prevent spawning and movement of carp downstream if the lakes spilled. Funding from the Tasmanian government allowed lake levels and flows to be modelled by the Hydroelectricity Commission. This resulted in alterations to the canal between the two lakes and the construction of levee banks which funnelled overflows from Lake Crescent down a screened channel. Subsequent construction works duplicated the outlet from Lake Crescent. Ongoing electrofishing surveys downstream in the River Clyde, beginning in February 1995, confirmed that the screens, water management and other works prevented virtually all escape of carp from the lakes for the duration of the Carp Management Program (Yick *et al.* 2021). The only escapee was an adult carp, found in the outlet to Lake Crescent during de-silting works in July 1995 (Inland Fisheries Service 2004).

FISHING THE CARP DOWN

The Tasmanian government committed more than $1 million in June 1995 to carry out the capital works described above, run public and angler awareness campaigns and establish a four-person team to catch and remove carp from the lakes (Diggle *et al.* 2012). Reducing the numbers of carp in the lakes, with a view to eradication over time, was the obvious next priority once the fish were contained. In reality, catching carp began pretty much from Day 1, and indeed physical removal turned out to the be the most effective method for reducing and eventually eradicating carp. But early on, given the desperate nature of the situation, a range of eradication methods was canvassed.

High on the list of methods were draining the lakes, poisoning using rotenone (derris dust) and actual physical removal (Diggle *et al.* 2012). The first option was quickly dismissed because of the topography of the lakes, the conservation significance of the Interlaken Lakeside Reserve and the flora and fauna inhabiting the lakes, and because the lakes supplied water for towns and farms downstream. As far as the second option was concerned, a company in the US that produced rotenone was consulted about the feasibility of the chemical's use. Although in theory rotenone could have been applied – at a cost of US$4.8 million in 1998 money – it would have required all the world's production of the chemical in that year, the lakes would have had to be drawn down, and there was no example of it ever having worked at that scale anywhere in the world (Wisniewski *et al.* 2015). A few other methods were considered and/or trialled before starting the physical removal of carp: a carp-specific poison (that would need to be developed); a 'suicide' gene that could be switched on if the population was swamped with genetically manipulated fish (also to be developed); spring viraemia of carp virus (not deployed because of questions about its efficacy in wild populations and possible effects on native fish); daughterless carp technology, which was being developed by CSIRO scientists (Thresher and Bax 2003) as a way of bringing about the production of only male offspring (yet to show feasibility and only in its infancy at the time) (Inland Fisheries Service 2004); and berley (pellets of corn and meal were trialled in an attempt to attract carp, but proved ineffective) (Diggle *et al.* 2012). The cyprinid herpesvirus (CyHV-3) was not at the time on anybody's radar (see Chapter 10).

Fig. 9.3. Seining for carp in Bullies Marsh, Lake Crescent. Source: Jonah Yick, Tasmanian Inland Fisheries Service.

As a critical part of the fishing down of carp, methodological studies and research into fundamental aspects of the carp populations in the lakes were initiated. These included: efficacy and selectivity of capture methods (Macdonald and Wisniewski 2003; Walker and Donkers 2011); estimates of population size and detection by CSIRO researchers and others (Donkers 2003b; Donkers *et al.* 2012; Furlan *et al.* 2019); age and growth of the population (Donkers 2004); and movement and habitat use in the lakes, especially in relation to spawning (Diggle *et al.* 2012; Taylor *et al.* 2012). As we saw in Chapter 7, the carp in Lake Crescent was genetically a mixture of Boolarra and koi (Davis *et al.* 1999), which suggested that the introduced carp had come from wild mainland populations (Inland Fisheries Service 2004).

In the early days, Carp Management Program officers were fishing by the seat of their pants, as it were, with limited knowledge of the population size, age structure or habitat use of carp in lakes Crescent and Sorell – in short, all the information that is needed to locate and make physical removal more effective (Fig. 9.3). So at first, they relied mostly on finding aggregations of carp visually and then using electrofishing and gill and seine nets to catch them (Diggle *et al.* 2012; Wisniewski *et al.* 2015). But lessons were learnt, methods were refined, and new and better ways of locating and catching fish were used. Length data, for example, assisted in choosing the best gear to use. The focus was on fishing-down carp in Lake Crescent, because survey work had indicated this was where the bulk of fish were (Yick, pers. comm.). Lake Sorell was fished concurrently, but not as systematically until much later.

At the start of the fishing-down campaign, surveys indicated that there were probably two cohorts, but possibly more, of carp in Lake Crescent, made up of 200 adults and 4,000 juveniles

(Inland Fisheries Service 2004). Some of the larger carp (>520 mm caudal fork length) may have been more than 10 years old. The estimated large number of small carp were likely from a successful 1992 spawning event. Two successful spawnings occurred in 1996, in spring and late summer, producing more than 3,000 recruits (Donkers 2004). In contrast, in 1995 Lake Sorell had far fewer carp, possibly only 30 in total, including adults and juveniles, although there were subsequent successful recruitment events in that lake (Donkers 2003b). Ageing of carp in the lakes using otoliths was problematic and inconsistent results were obtained (Donkers 2004). Thus, growth rates were not considered reliable and there was uncertainty as to how old some fish actually were.

When numbers of carp were large, seine netting was particularly effective. But declining numbers brought the problem of how to locate the fewer remaining individuals (Fig. 9.4A).

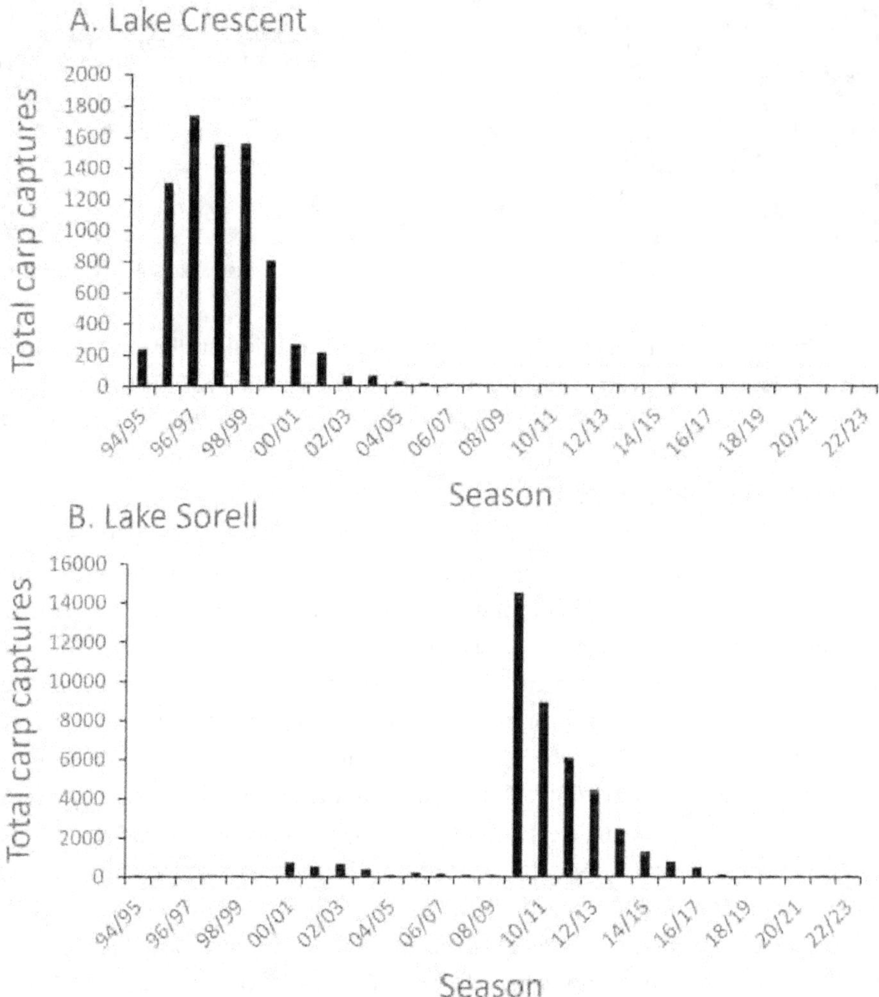

Fig. 9.4. Number of carp caught in (A) Lake Crescent and (B) Lake Sorell, 1995–2023. Sources: Yick *et al.* (2021), Inland Fisheries Service (2023).

Declining catch-per-unit-effort was clearly a good thing as it showed that the methods were working and fish were not being replaced with new recruits (see below), so there was cause for celebration that physical removal was working. But more was needed.

In 1997, biotelemetry was first used as a way of tracking carp movements and locating aggregations in a more efficient way (Macdonald and Wisniewski 2003). It took advantage of the carp's propensity to gather in groups when feeding and breeding. Initially in early March, nine 'Judas' or 'transmitter' males with implanted transmitters were released into Lake Crescent and followed daily. Females were not used, to avoid replacing carp in the lake that might contribute young to the next generation. A month later, 202 carp were caught in an aggregation, having been betrayed by one of their own. The accumulation of movement and aggregation data over time allowed Carp Management Program scientists to predict, based mostly on lake level and temperature, where aggregations would most likely occur (Diggle *et al.* 2012). From then on, this method was routinely used and it proved very effective (Fig. 9.4A). Strategically, the Carp Management Program moved from containment, then to control and on to eradication.

CONTRACEPTION

A critical step to eradication, beyond fishing down the contained carp, was preventing the production of future generations. Several ways of doing this were applied in tandem with reducing carp numbers. Fortunately, the region's climate and the carp's behaviour were on the side of the Carp Management Program team. The cool temperatures on the Central Plateau and in the lakes meant that carp grew slowly and reached maturity relatively late, thus intergenerational times were longer than in warmer regions of the world (Yick *et al.* 2021). Furthermore, the fact that carp congregates – and as we have described in several chapters, does so conspicuously – makes the fish easier to spot and target.

On-ground works to prevent spawning began in 2004 in Lake Crescent, when shallow wetland sections of the lake with large amounts of aquatic plants were fenced off with chicken wire (Diggle *et al.* 2012). Lockable steel traps were constructed and placed in locations where carp would likely attempt to get into the wetlands (Fig. 9.5). These were very effective during the spawning time of carp as temperatures and water levels rose, and caught more than half of the total number of fish captured in the 2004/05 spawning season. Fences, however, did not prevent access to the wetland spawning areas when lake levels were high. The effectiveness of the recruitment-prevention methods was increased through the removal of eggs when located, the use of hydrated lime to poison the aquatic plant beds on which eggs were attached (Diggle *et al.* 2012), and through dropping lake water levels several weeks after carp embryos hatched so as to poison the aggregations in now-isolated channels and depressions using rotenone (Yick *et al.* 2021).

Annual recruitment surveys were initiated in 2004 to guide subsequent actions, such as targeting aggregations of juvenile carp with fyke nets, small-mesh gill nets and electrofishing (Diggle *et al.* 2012). These surveys are ongoing.

Fig. 9.5. (A) Setting up a barrier trap during autumn. (B) The same trap in operation in spring. Source: Jonah Yick, Tasmanian Inland Fisheries Service.

ON THE TAIL OF THE LAST CARP IN LAKE CRESCENT

Apart from helping Carp Management Program officers to locate aggregations of carp, the biotelemetry work showed that some carp were particularly smart at avoiding capture in nets or traps (Diggle *et al.* 2012). But they couldn't resist the urge to reproduce, which provided another opportunity to reduce numbers. In 2005, female carp which had been injected with

Fig. 9.6. Chris Wisniewski holding the the last carp caught in Lake Crescent, December 2007. Source: Jonah Yick, Tasmanian Inland Fisheries Service.

pituitary extract from mature fish were put in mesh bags behind the steel traps described above, to attract males (Patil and Wisniewski 2011). A plume of irresistible odour emanated from these carp 'sirens' and led to the capture of nine males, including a transmitter male (Yick *et al.* 2021). This was followed with the implantation of pumps containing prostaglandin F2, which is known to attract carp (Sorensen *et al.* 2019). But the pituitary method seems to work best.

By 2007, the Carp Management Program team knew there was only a handful of carp left in Lake Crescent. Jonah Yick told us that, in that year, transmitter males led the team to 15 separate aggregations, 14 of which yielded only transmitter carp (Yick *et al.* 2021). The team was on the cusp of complete eradication. Some rain and warm humid weather resulted in two transmitter carp moving into shallow water in December of that year. Three fish were caught: two transmitter males and one female carp (Fig. 9.6). This female was the last carp pulled out of Lake Crescent. Successful eradication of carp from Lake Crescent was officially announced in 2009.

Although the battle for Lake Crescent was won, the war with carp in Tasmania was not over.

A CRISIS IN LAKE SORELL

Methods developed, trialled and refined in the smaller Lake Crescent were subsequently used in the larger Lake Sorell. Modifications were further refined to take into account lessons learnt and the different size, layout and behaviour of the lake (Yick, pers. comm.). For example, polypropylene nets were used as a barrier instead of chicken wire fences, and gill nets were added behind this barrier during the spawning season (Yick *et al.* 2021; Inland Fisheries Service 2023). Additionally, whereas multiple gill nets were used in Lake Crescent when carp aggregations were found, in Lake Sorell trammel nets (gill-like nets that have three layers of different-sized mesh) were used, because they can catch multiple size classes of fish at the one time (Yick, pers. comm.).

But soon after the last carp was caught in Lake Crescent and numbers of carp were very low in Lake Sorell, disaster struck. For a time, it seemed that there was a very real chance that carp had got away from the Carp Management Program team and all their hard work over the last 14 years would be undone.

By the 2008/09 season, it was estimated that only 38 or 39 adult carp remained in Lake Sorell (Fig. 9.4B) (Inland Fisheries Service 2010). The preceding decade or so had been a particularly dry one – the Millennium Drought – which meant that lake levels were generally lower than the long-term average. The drought broke in late winter/early spring of 2009 with a vengeance, and lake levels rose dramatically at the same time as the weather warmed. Although barrier nets and traps were in place in Lake Sorell, there was not enough netting to prevent fish from getting to Silver Plains Marsh. Fish spawned and eggs were found in the wetlands. The population of carp in Lake Sorell jumped by an estimated 50,000 recruits.

This was devastating news. But the IFS held its nerve (John Diggle and Chris Wisniewski, pers. comm.). A workshop was held and a review was conducted in early 2010, resulting in recommendations to the Tasmanian government. As a result, extra funding was obtained, which the federal government later matched (Inland Fisheries Service 2010). What added insult to injury was the realisation that non-sterile male transmitter carp had contributed to the spawning and subsequent recruitment. It was decided to replace the adult male transmitter carp with juvenile transmitter carp, and investigations were initiated into using only sterile transmitter males as Judas carp (Inland Fisheries Service 2011). More staff were put on, and works were carried out to prevent carp moving back into Lake Crescent –a real threat, as juveniles were found on screens in the channel linking the two lakes. Fishing-down efforts were scaled up dramatically, and methods were adapted to target juveniles. Thousands were caught then and for the next few years. The urgency of preventing future similar spawning and recruitment events galvanised efforts to prevent potential spawners from getting to spawning habitat, which is where the gill nets placed behind the barrier nets came in. All of these efforts paid off. As Fig. 9.4 clearly shows, the numbers of carp declined consistently over subsequent years as the 2009/10 recruitment cohort was fished out and further spawning was prevented.

The efforts to eradicate carp from Lake Sorell were given a fortuitous boost by the discovery in 2014 of males with a gonadal abnormality – 'jelly gonad syndrome' – that can make the male sterile (Inland Fisheries Service 2016). One in four males in Lake Sorell had the condition early in the 2015/16 season, and by the latter half of 2016 the prevalence had increased to one in every 2.4 males. A project by then University of Tasmania PhD student Raihan Mahmud was carried out to understand more about this disease (Mahmud *et al.* 2020). Overall, it considerably reduced the number of potential breeding males.

Fig. 9.7. Jonah Yick holding the last carp caught in Lake Sorell, November 2022. Source: Jonah Yick, Tasmanian Inland Fisheries Service.

Through sound decisions, extra funding and an enormous amount of on-ground work by the Carp Management Team, the crisis had been averted. Within a couple of years, and with no spawning in Lake Sorell in most years, Carp Management Program Annual Reports could offer more up-beat news. The 2009 cohort of carp continued to be fished down and the last carp was caught in Lake Sorell in November 2022 (Fig. 9.7). As we write this in early 2025, carp had been declared functionally eradicated from Tasmania by the IFS. Recruitment monitoring has stopped in Lake Crescent but will continue for a little while longer in Lake Sorell just to make sure (Yick, pers. comm.). Environmental DNA studies (eDNA) have also been used to ensure the absence of carp (Furlan *et al.* 2019). In all, 7,797 carp were removed from Lake Crescent in 12 years and 41,504 from Lake Sorell in 28 years.

*****

Success stories like the Tasmanian one are rare. A big shout-out to all the people involved in the incredible work of learning about, controlling and ultimately getting rid of carp from Tasmania, especially Wayne Fulton, Andrew Sanger, John Diggle, Paul Donkers, Chris Wisniewski, Jonah Yick, Jawahar Patil and the many others who dedicated time and effort, under often very difficult conditions, to deal with such a complex and demanding environmental problem. The members of various committees were also instrumental in advising on the unfolding and dynamic carp dilemma. The Tasmanian government deserves an honourable mention too, for realising the need for substantial and ongoing funding and support to tackle such a huge problem. The Tasmanian experience is a reason to celebrate and there are lessons that can be learnt. Some obvious ones are the need for adequate funding, a multi-pronged approach, fast responses to changing situations and a lot of dedicated and hard work by many people. A less obvious lesson perhaps is the importance of institutional knowledge. Despite a huge turnover of employees during the almost 30 years of the Carp Eradication Program, the IFS maintained a continuity of experience at the senior level (John Diggle and Chris Wisniewski, pers. comm.), which meant that knowledge gained was not lost over time, and success was built on success. Longevity of experience and knowledge is rare with environmental problems such as the one in Tasmania, and its critical importance should be recognised. Of course, the situation in the Murray–Darling Basin, and in any of the river systems elsewhere on the Australian mainland where carp occur, is very, very different from that of lakes Crescent and Sorell. So, let's hang on to the feeling of optimism that the Tasmanian experience has given us while we can, while guarding against complacency that such incursions could never happen again. We need all the optimism we can get: the next chapter talks about the true scale of carp in Australia, and what can be done about it.

10

Carp control

Even if it does require a version of a venereal disease to deal with it. If that's what's required, that's what's required! (Barnaby Joyce, Honourable Member for New England, House of Representatives, May 2016).

ANYBODY WHO HAS SPENT TIME WANDERING AROUND GIPPSLAND OR THE Murray–Darling Basin (MDB) during the last 40 or so years cannot but have noticed carp swimming in huge numbers in virtually every water body that they come across. The fish is seemingly everywhere. Carp are there in spring, swimming in groups, splashing around noisily in an apparent orgy of spawning. A month or so later they are there as new recruits that wriggle like vast greeny-gold organic oil slicks, moving from spawning to nursery grounds to grow rapidly into another generation of big plump fish. And they are there during the rest of the year, rising to the surface, their rubbery lips and whiskers unmistakable even to the casual observer, occasionally jumping clear of the water as they try to catch an insect. Or seemingly, jumping just for the hell of it.

As we described in Chapter 8, Ivor Stuart and colleagues estimated that in an average climatic year, Australia supported almost 200 million carp (with error estimates included, the number was somewhere between 106 million and 358 million). When conditions were especially wet, there could be a whopping 358 million carp (lower–upper limit = 179–685) swimming around in our waterways. If that was not sufficiently sobering, we outlined the possible effects of carp in Australian freshwaters. Multiply these effects on native animals and plants and the water quality of our river floodplain ecosystems by 100 million or so, and it is enough to put you off fish and chips for life.

It is not surprising that the total numbers of carp – not just their lips – are a bit rubbery, given they are based on models, and models are only approximations of reality. But it is also not surprising that populations of carp rise dramatically when dry times become wet times, considering what we know of the species' biology (Chapter 6). Our knowledge is based not on models but on actual data collected over 30 years (Forsyth *et al.* 2013). Some studies suggest that carp do not comprise as high a percentage of the fish biomass, in at least the New South Wales part of the MDB (Schilling *et al.* 2024), as others have suggested (up to 90% is commonly mentioned: Koehn 2004; McColl and Sunarto 2020). But even a median of 57% of the biomass of all fish in a catchment is a startling figure for any fish, native or non-native.

As a result of its spread, fishy dominance and sheer numbers, carp has understandably been a major concern of the general public, conservationists, natural resource managers, ecologists and politicians for decades. The initial Victorian carp-kill program in the 1960s aimed to eradicate Boolarra carp before it spread further, but it failed (Chapter 5). However, the program provided valuable lessons and a stark reminder of how hard it is to exterminate an invasive species once it has begun to spread. We now recognise that legislation is critical to prevent invasions in the first place, and that poisons are effective only in small isolated water bodies and can harm non-target species. We have learnt that getting the public involved is absolutely crucial if control, management and potentially eradication are to have a chance of success. And we have learnt that insights into carp biology, which was pivotal to the species' eradication in Tasmania, help with control strategies and form much of the foundations of pest control strategies employed today.

This chapter focuses on approaches to the control and management of carp in Australia since the Boolarra carp got away from Victorian fisheries authorities in the early 1960s. We first outline some pest management principles that are key when considering how to deal with species like carp. We then briefly describe the history of carp control management strategies, before listing the numerous methods that have been, and are yet to be, deployed in controlling, and perhaps – although it is unlikely – eradicating carp from our freshwaters, together with their advantages and disadvantages. We include a couple of illustrations of what we view as particularly interesting aspects of the story of management of carp in Australia: the Williams separation cage, Keith Bell and his commercialisation of carp, and finally the controversy surrounding the potential release of the cyprinid or koi herpesvirus.

PEST MANAGEMENT PRINCIPLES

Eradication of carp from Tasmania was achieved, but it took close to 30 years. And that was in two relatively small lakes – closed systems, within which carp could be contained. Controlling carp populations on mainland Australia, across a vast network of rivers, lakes and floodplain water bodies, is a far, far more complex proposition. After 60 years of work we are discovering, as they have in North American and New Zealand, that control and management are challenging enough, and the likelihood of eradication is highly improbable.

Mary Bomford and Richard Tilzey first presented 'Pest management principles for European carp' at the Controlling Carp: Exploring Options for Australia workshop in 1996 (Bomford and Tilzey 1997), and the Centre for Invasive Species Solutions (pestsmart.org.au) further developed those principles to cover vertebrate pests in Australia more generally. The full details of each principle can be found at the Centre for Invasive Species Solutions website factsheet for carp (Centre for Invasive Species Solutions 2014). Some key points of the principles are:

1. an animal is a pest only where it is not valued; for example, carp may be a pest in Australia but not in Austria;
2. people have different attitudes towards pest animals, which has a big bearing on how management programs should be implemented;

3. pest animals are unlikely ever to be eradicated, and so preventing them entering in the first place is the best course of action;

4. management of a pest should be based on cost/benefit assessments and should focus on minimising the damage that the pest causes;

5. an understanding of integrated systems of community structure, habitat, movement, food webs, population dynamics and genetics is needed for effective pest management;

6. we will never know everything about the ecosystems in which the pest lives, and so we need to learn and adapt as we go; and

7. it is vital to use monitoring and evaluation techniques that will let us determine if we have met our management goals – otherwise, we don't know if we have been successful or not.

Most recent nationally coordinated carp control strategies have considered many, if not all, of these principles from the outset.

In the case of carp, apart from the really early days when fish had only just been detected, most management strategies have aimed at controlling carp by reducing their numbers and preventing them from getting into new areas or into areas that might enhance their spawning and subsequent recruitment (Stuart *et al.* 2021). More drastic control methods have been on the agenda of most incarnations of nationally coordinated strategies (Fig. 10.1). In the next section, we briefly summarise the strategies, summits, workshops and organisations involved in carp control and management, before listing the methods that have been involved. We include details of a few methods that in our view deserve special consideration.

AUSTRALIA'S NATIONAL STRATEGY FOR CARP MANAGEMENT

The first carp programs were state-based, and included a mix of research on the biology of carp, experiments on carp's effects on aquatic ecosystems and assessments on what options existed for control and even eradication (Fig. 10.1). By the early 1990s – although many had recognised the scale of the problem almost two decades previously (and for Alf Butcher and Jim Wharton, even earlier than that) – a more national approach to carp was needed. Public pressure was largely responsible for this more systematic and unified approach, which culminated in two carp summits, one in Wagga Wagga in 1994 and one in Renmark in 1995. The foundation of the National Carp Task Force (NCTF, run through the Murray–Darling Association), which included a range of stakeholders, followed shortly after (Barrett 2003). Through lobbying by the NCTF, the federal government formed the Carp Control Coordinating Group (CCCG) in 1997 in recognition of the need for a national approach to carp control and management. Various stakeholders were part of the CCCG, mostly representatives from federal and state natural resource agencies, but also conservation groups and scientists. The CCCG reported to several Ministerial Councils associated with water quality, water delivery, agriculture, conservation and fisheries (Barrett 2003). But at that stage, no Indigenous representatives were included.

The NCTF and CCCG had milestones that included coordinating state and federal agencies, research and community engagement, all aimed at mitigating the impacts of carp in the MDB. Key initiatives from the CCCG included developing the National Management Strategy for Carp Control (NMSCC) and a number of documents, which, together with the report authored

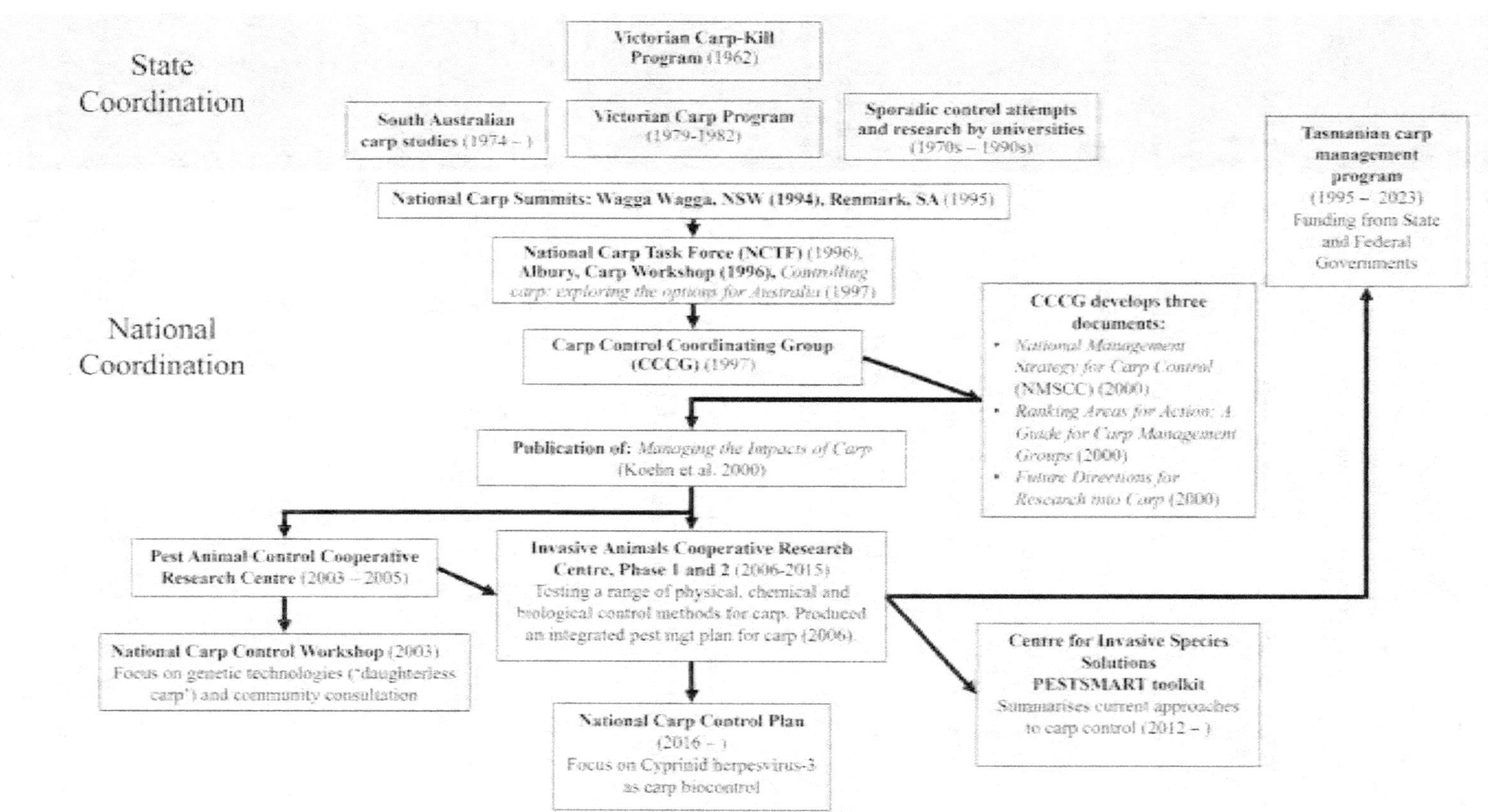

Fig. 10.1. Australia's frameworks for managing common carp, 1962–present day.

by John Koehn, Andrea Brumley and Peter Gehrke, *Managing the Impacts of Carp* (Koehn *et al.* 2000), have been instrumental in guiding the control and management of carp ever since.

Managing the Impacts of Carp, especially, integrated what we knew at that time about carp biology, control techniques, economic and social impacts, assessments of carp as a commercialised product, humane treatment and adaptive management. The authors interviewed, and reported the views of, a range of stakeholders such as government bodies, researchers, Indigenous groups, recreational fishers and community groups. The report laid the foundations for later carp programs included in the Pest Animal Control Cooperative Research Centre (2003–2005) and the Invasive Animals Cooperative Research Centres (2006–2015). These organisations were instrumental in formulating and testing the latest integrated pest control technologies – including physical, chemical and biological control options – and collaborated with international experts. They also commissioned projects, like CARPSIM, which allowed various management approaches to be tested virtually through modelling of carp populations (Brown and Walker 2004; Brown and Gilligan 2014). The Pest Animal Control CRC first proposed investigating the potential of daughterless carp technology and the Invasive Animals CRC was the group that began investigations into the use of cyprinid herpesvirus-3 (CyHV-3) or koi herpesvirus (KHV) as part of its integrated pest management plan for carp (Fulton 2006). Later in this chapter we will describe in detail the virus and issues surrounding its use.

METHODS FOR THE CONTROL AND MANAGEMENT OF CARP

Some carp control methods and research go back many decades, as far as the 1960s. But over the past 20 years or so, the Invasive Animals CRC has been the driving force behind developing and trialling a range of methods and it has guided scientific research. It has focused on developing effective strategies for Australia and other regions where carp are considered a problem, in particular North America. The strategies have been based on the pest management principles, and no method is perfect. All have advantages and disadvantages and all have trade-offs in effectiveness, ease of use, size- or species-specificity, impacts on non-target organisms and cost. Some are more attractive to, and even involve, the general public. Some are decidedly unattractive to the general public, and require strong and sustained engagement. In the early days, Indigenous groups were rarely included or consulted, but over the years this has changed. The National Carp Control Plan consulted First Nations peoples on their views and priorities, and Aboriginal Ranger groups have been involved in the physical removal of carp in some parts of the MDB (Sanders and Morris 2018). We hope that the involvement of, and consultation with, First Nations communities develops further in the future, considering their profound knowledge of native fish and the ecosystems in which they live.

The methods we present in Table 10.1 are not exhaustive. We do not, for example, include genetic isolation as a method of reducing the viability of carp populations over time (Brown 1980), as in recent decades it has not been taken up. Nor do we include immune contraception, for the same reason (Hinds and Pech 1997). We present the methods that have been most used or proposed, and briefly discuss the merits of each. Broadly, they can be classified into physical

Table 10.1. Advantages and disadvantages of carp control and management methods that have been used or proposed in Australia.

Type of control	Method	Advantages	Disadvantages	References
Physical removal	Electrofishing Seines, fykes, gill or trammel nets Trapping 'Judas' carp Commercial fishing	At high densities, removes large numbers rapidly Species-specific Can be environmentally sensitive Good for shallow lakes and dams Fish can be used commercially Methods well tested	Effective only locally and in the short term Not good in deep water, high turbidity or fast current Poor efficiency in open systems with structure and flow Avoidance behaviour can develop Labour-intensive and costly Low consumer demand for carp Bycatch can be unacceptable Size-selective Risk of new populations if commercially profitable	(Thresher 1997; Koehn *et al.* 2000; Brown and Walker 2004; Gilligan *et al.* 2005, 2010; Graham *et al.* 2005; Weber *et al.* 2011; Colvin *et al.* 2012; Centre for Invasive Species Solutions 2014; Simonson *et al.* 2022)
Recreational fishing	Community fishing competitions, musters, fish-outs Rod and reel and bow fishing Sometimes prizes offered	At high densities, removes large numbers rapidly Non-target species released. Community-driven, popular, family friendly events Raise awareness and engage with community Low-cost Fundraising for other river restoration or native fish stocking activities	Effective only locally and in the short term Size-specific Lower rates of capture than commercial harvesting Rarely used commercially except as fertiliser	(Koehn *et al.* 2000; Brown and Walker 2004; Graham *et al.* 2005; Gehrke *et al.* 2010; Norris 2011; Norris *et al.* 2013, 2014)

Type of control	Method	Advantages	Disadvantages	References
Movement traps	Williams carp separation cage	Can be installed widely Removes large numbers May eliminate upstream populations Suitable for existing or newly designed fishways Supports commercial harvesting and composting Returns can exceed set-up costs Provides continuous supply Collaboration needed to maintain the cage Minimal bycatch May be tool for population control or as part of broader programs	Size-specific (generally captures carp >250 mm) Relies on movement and suitable water levels Over-removal of jumpers may select for non-jumping adaptations, reducing long-term efficiency Variable site effectiveness Maintenance is labour-intensive Minimum annual catch of 40–50 tonnes needed for commercial viability Limited feasibility in areas lacking carp disposal options	(Stuart et al. 2006; Thwaites et al. 2010; Koehn et al. 2018)
Poisons and explosives	Poisons, e.g. rotenone Chemicals that change pH, e.g. lime Explosives, e.g. dynamite	At high densities, removes large numbers rapidly Good in closed systems Good for aggregations, e.g. during spawning	Not size- or species-specific Limited to small closed systems Time- and labour-intensive Costly Illegal in some jurisdictions Poor community acceptance Fish cannot be used commercially	(Wharton 1971; Hume et al. 1983; Fajt and Grizzle 1998; Koehn et al. 2000; Gehrke 2001; Gilligan et al. 2005; Rayner and Creese 2006; Dalu et al. 2020)
Draining, habitat and flow	Wetland draining Habitat modification Flow alteration	Good for small closed systems with water control mechanisms Species-specific if native species relocated Elimination of fish in impoundments during spawning Can be done regularly, e.g. to prevent access to spawning habitat	Limited to small closed systems May be difficult during high-flow events Length of spawning time may make method prohibitive Habitat rehabilitation can take decades to show response Establishing cause and effect problematic	(Koehn et al. 2000; Sivakumaran et al. 2003; Nicol et al. 2004; Stuart and Jones 2006b; Weber and Brown 2009; Koehn et al. 2018; Todd et al. 2024)

Table 10.1. *continued*

Type of control	Method	Advantages	Disadvantages	References
Carp exclusion and containment devices	Physical (e.g. mesh screens) and behavioural (e.g. sound, air bubbles, electrical current) barriers reduce spread of fish Can be permanent, temporary or seasonal	Can prevent mature carp accessing spawning grounds Remove large numbers locally, especially during breeding Good at isolating from other water bodies Relatively inexpensive Mobile barriers deployed rapidly	Size-specific, mostly excluding large carp May exclude native fish Reduced effectiveness during floods Maintenance of equipment is labour-intensive (although self-cleaning screens have been developed) Complete exclusion rarely achieved Fish may learn to pass behavioural barriers Requires knowledge of carp biology	(Gilligan *et al.* 2005; Turnpenny and O'Keeffe 2005; Hillyard *et al.* 2010; Hillyard 2011; Keller 2014; Thwaites and Cheshire 2016; Thwaites and Schmarr 2019; Smith-Root 2025)
Attractants and repellents	Chemicals (e.g. food, wetland plants) Pheromones Sound	Species- and life stage-specific Various flow conditions Tiny amounts often needed Best used with other methods, e.g. traps and cages	Methods still under development Pheromone supply must be regularly replenished Risk assessments needed for approval Fish may learn and adapt to avoid Traps or cages need to be regularly removed	(Popper and Carlson 1998; Elkins *et al.* 2009; Lim and Sorensen 2012; Sloan *et al.* 2013, 2019; Carl *et al.* 2016; Bzonek 2017; Bzonek *et al.* 2020; Elkins 2021; Flores Martin *et al.* 2021; Ghosal *et al.* 2022; Bullers 2024; Kasumyan *et al.* 2024)
Predation	Fish as predators Birds as predators Zooplankton as predators of larvae	Native fish, e.g. Murray cod and golden perch, are known to eat carp at times Native birds, e.g. pelicans and cormorants, are known to eat carp at times Cyclopoid copepods are known to eat freshwater fish larvae Different life stages eaten Native fish stocking currently underway Stocking relatively cheap and can be done by community Good for range of systems	Predation not sufficient to reduce populations Juvenile carp located where predators absent Predation not limited to carp Predation limited by gape of predator Overstocking predators may lead to collapse if insufficient prey Predators may move from carp areas Difficult to monitor effects	(Davis 1959; Fabian 1960; Koehn *et al.* 2000; Ebner 2006; Bajer and Sorensen 2010; Bajer *et al.* 2012; Doyle 2012; Poole and Bajer 2019)

Type of control	Method	Advantages	Disadvantages	References
Biological control: 1) sex biasing technology	Daughterless carp: genetic modification to produce only male offspring Trojan fish (e.g. Trojan Y chromosome)	Method could in theory eradicate carp Species-specific	Expensive Still under development – no proven success for any fish to date Effectiveness depends on heritability, fitness, population size, numbers released Requires integration with other methods Problems with public acceptance May take 100 years to work Challenging to implement May negatively affect koi and commercial industries	(Koehn *et al.* 2000; Thresher and Bax 2003; Brown and Walker 2004; Brown *et al.* 2005; Fisher and Cribb 2005; Thresher 2008; Centre for Invasive Species Solutions 2012; Hayes *et al.* 2014; Teem and Gutierrez 2014; Teem *et al.* 2014; Thresher *et al.* 2014)
Biological control: 2) viruses/ diseases	Spring viraemia carp virus Cyprinid herpesvirus (CyHV-3) Jelly gonad condition	Potential to cause mass mortalities or infertilities Species-specific Good under high-density, closed-system conditions Jelly gonad condition a factor in eradication of carp from Lake Sorell in Tasmania	Can be size-specific: juveniles less susceptible Upper and lower effective range of virus Hybridisation with goldfish reduces virus effectiveness Detection of resistance to and latency of virus may reduce effectiveness Virus less effective on European strains At best 80% effective. Likely 40–60% Public consultation essential Public, private and scientific resistance to release of biological agent a risk Mass mortalities require clean-up and may cause water quality degradation Impact of immune-compromised carp on ecosystem not fully considered	(Brown 1980; Baxter *et al.* 2005; McColl *et al.* 2016, 2017; Cyr *et al.* 2017; Thresher *et al.* 2018; Boutier *et al.* 2019; Kopf *et al.* 2019a; Lin and Landos 2019; Joehnk *et al.* 2020; Mahmud *et al.* 2020; McColl and Sunarto 2020; Graham *et al.* 2021; Mintram *et al.* 2021; Samsing *et al.* 2021; Mankad *et al.* 2022; Inland Fisheries Service 2023; Wang *et al.* 2024)

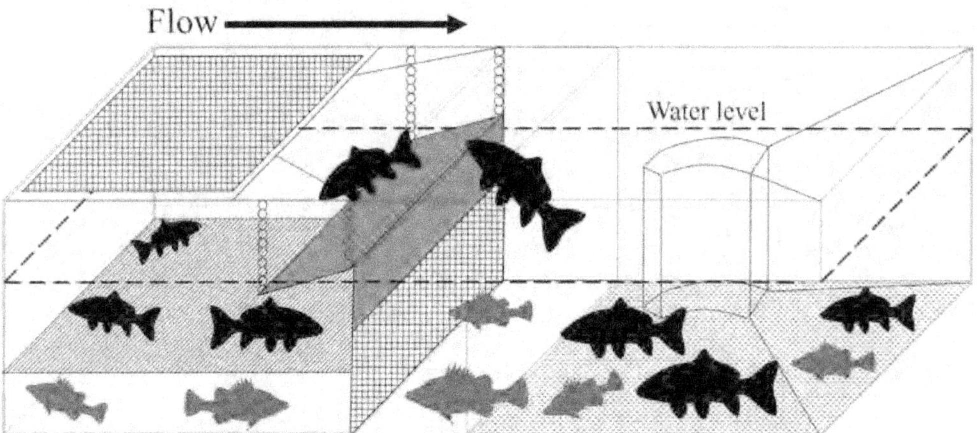

Fig. 10.2. The Williams carp separation cage. In the figure, carp silhouettes have smooth dorsal fins; golden perch have spiny first dorsal fins. Source: Redrawn after Stuart *et al.* (2006).

removal, recreational fishing, trapping of moving fish, the use of poisons or explosives, draining water bodies or altering habitat or flow, exclusion or containment methods, attractants or repellents, predation and biological controls. References for each type of control are given in the right-hand column of Table 10.1.

Physical removal has been employed broadly in Australia and overseas. It can involve active (e.g. electrofishing) and passive (e.g. fyke netting) methods (Table 10.1). These methods are useful when densities are high and in environments that are free from structure and are still or slow-flowing. But they are mostly effective at small scales and in the short term, unless employed consistently and intensively. Commercial fishing has the potential to remove large numbers of carp if employed consistently and intensively. Recreational fishing, a form of physical removal, sometimes occurs through the action of individuals and sometimes through organised competitions, musters or fish-outs. These have the benefit of community engagement and awareness, can be effective at small scales and mostly cost little, but they are not the answer for wholesale control.

Movement traps are an innovative way to catch large numbers of carp. A notable one is the Williams carp separation cage (Fig. 10.2) (Crook and Gillanders 2006; Stuart and Conallin 2018). This method was developed through observations of the movement behaviour of carp – or attempted movement – through fishways. At some fishways, fish are prevented from moving upstream so that they can be counted and carp separated from native fishes. Alan Williams, a weir keeper at Torrumbarry Weir, noticed in the early 2000s that trapped carp typically tried to escape by jumping. Native fish don't jump when trapped. That gave him the idea of building a cage to separate carp from native fishes, by trapping carp that jumped in a separate compartment. Non-jumping native fishes would be able to continue their journey upstream, while carp could be removed and used commercially. Over 11 years at Lock 1 on the Lower Murray River, almost 300,000 carp were caught in such a cage (Stuart and Conallin 2018). That amounted to a maximum of 5 tonnes/day. The fish sold commercially made almost A$1 million, which well-and-truly covered the cost of the cage. The bycatch of native fish was negligible.

Together with the effects of commercial fishing and low flows, the Williams cage contributed to reduction in carp numbers below Lock 1 (Stuart and Conallin 2018).

Poisons have been used at times to kill carp, almost always in small closed systems such as lakes or artificial water bodies. But, as the Tasmanian experience showed, the amount of poison required for relatively large lakes can be enormous, prohibitively expensive and impractical to acquire and distribute (Diggle *et al.* 2012). Explosives have been tried at times, such as during the Victorian Carp Program, but have proven typically ineffective (Hume *et al.* 1983).

Altering or enhancing habitat or flow or even draining carp-infested wetlands have been considered as approaches for carp control. There is a certain logic in that if an ecosystem is intact, native fish are abundant, there is copious habitat and flows are natural, carp may struggle to establish. Modelling by Charles Todd and colleagues suggested that, because carp require inundated vegetation for successful spawning and recruitment, it is not surprising that natural floods and prolonged droughts greatly influence carp population sizes (Todd *et al.* 2024). Furthermore, artificial flooding of floodplain habitats can lead to an increase in carp populations. River managers may aim to improve floodplain habitats through environmental watering, but that may come at the cost of more carp. In some cases, wetlands can be drained, carp removed and native fish relocated, but in most cases draining wetlands is impractical. It is better, if possible, to exclude carp from water bodies in the first place by using physical or behavioural barriers. The advantage of such devices is that they can be deployed seasonally and installed and uninstalled relatively quickly. Carp, being relatively bright fish, may learn how to avoid or become accustomed to behavioural barriers.

Attractants and repellents have been increasingly trialled in recent years. These include pheromones, which we mentioned in the context of the Tasmanian experience. They have the potential to be species- and life stage-specific, but should normally be used alongside other methods which trap the attracted or repelled fish.

For a time, there was a view that if native fish capable of eating carp were in high enough densities, predation could be a way of controlling carp numbers. This was the basis of Katie Doyle's PhD work (Doyle 2012). Unfortunately, her results overall did not support this idea, because the size of carp as potential prey was limited by the gape of predators, which meant only juvenile carp were likely to be vulnerable; the habitat of prey and predators often did not overlap sufficiently; and if predators were stocked, they consumed native fish as well as carp, which could be problematic in some circumstances.

In the last 20 years or so, biological control of carp has been firmly on the agenda. Consideration of viruses has been around since the Victorian Carp Program and the spring viremia of carp virus (Hume *et al.* 1983); we will discuss the latest virus proposal at length below. The daughterless carp and Trojan Y chromosome strategies – where genes are altered in such a way that carp produce only males, with female genes being lethal, ultimately leading to local extinction of the population – were very attractive proposals that got serious traction in the early 2000s. They promised eradication of carp – if they were effective. However, the strategies have been a long time in development and have not been proven successful for carp to date. The effectiveness relies on a number of factors, including heritability of the genes, and they were always considered in concert with other control methods, such as the use of a virus. Public acceptability was an ongoing issue, considering the large numbers of genetically modified carp that would need to be released. And, since carp live for decades, in all likelihood, even if effective the strategies would take perhaps 100 years to completely eradicate carp.

Commercialising carp

Carp control and management requires a lot of money, sometimes millions of dollars. A commonly suggested alternative approach reframes carp from a valueless pest to a valuable resource through commercialisation of the fish. In this way, harvesting carp can have financial benefits while simultaneously reducing carp numbers. Since the spread of carp in the 1970s, the debate over carp commercialisation has been ongoing, with supporters highlighting benefits and critics warning about the risks (Koehn *et al.* 2000). It is not our intention to advocate the commercialisation of carp. We simply describe what has occurred in this area for many decades, but in the last 40 or so years has ramped up substantially. Concern has been expressed that commercialising carp risks economic reliance on the continued presence of the species (Koehn *et al.* 2000). Placing a monetary value on carp could be an incentive to maintain populations.

For most of the community, the enormous numbers of carp in the MDB are annoying, frustrating and get in the way of a good day's fishing. But some canny entrepreneurs have seen that carp can bring economic, environmental and social opportunities. Commercialising carp products can generate income and create jobs in fishing, processing and related industries, especially in regional areas, if markets are established (Jones 1995). Government assistance or harvesting and processing incentives can further support the industry (Koehn *et al.* 2000). This can be especially useful if harvesting doubles as a control method and recoups at least some of the costs. An example is the Williams carp separation cage.

Efforts to commercialise carp, by the likes of Keith and Cate Bell (K & C Fisheries) and Tracey and Greg Hill (Coorong Wild Seafood), have shown the species' versatility in providing products that people want, both within and outside Australia. Properly processed, carp has shown that it can be a 'no waste' species: all parts of the fish can be used and marketed. Globally, carp is a widely consumed and nutritious fish, valued for its high protein, phosphorus, vitamin B12, beneficial fatty acids and antioxidant content (Grand View Research 2024). The high protein content makes carp an alternative source in comparison to meat or other, non-sustainable fishes, particularly in regions with declining or over-exploited fisheries (https://www.abc.net.au/news/2022-08-14/carp-eel-could-be-budget-friendly-fish-as-cost-of-seafood-rises/101327174).

Although commercial fishing of carp has been around for a long time, early markets were mostly for crayfish bait and petfood. Carp are still sold for this market, but more recently commercial fishers, like K & C Fisheries and Coorong Wild Seafood, shifted focus and adapted their gear from general commercial fishing to specialise on carp. Despite its limited popularity, carp is now sold fresh for human consumption at fish markets in Sydney, Melbourne and Adelaide and at many smaller markets (including Albury), from areas like the Coorong, Lake Alexandrina and Gippsland Lakes, typically whole and on ice. While carp stocks could potentially support high annual harvest rates (Graham *et al.* 2005), low demand and fluctuating prices mean this resource is underutilised (Wilson 1998). Tastings and culinary demonstrations are used to improve public perceptions of carp and expand its market as an edible fish.

Carp recipes and food products shared through newspapers, fishing magazines and recipe books highlight carp as an environmental issue while encouraging communities to engage in pest control through consumption (Easton and Elder 1997). Businesses like Coorong Wild Seafood and Yabby City (run by the late Henry Jones) in South Australia have promoted carp in restaurants, with chefs such as Mark Best and Maggie Beer endorsing carp dishes (Berlage 2022). The recent ABC documentary *Eat the Invaders* further raised awareness of carp's potential as a food source (Armstrong 2025).

In 2003 K & C Fisheries Global developed a process for the production of high-quality carp roe for human consumption (https://www.kcfisheriesglobal.com/). Some fish sauce is produced from carp. In Eastern Europe, carp milt (sperm-containing fluid) is a traditional ingredient of Christmas dishes, including soups and fried or poached preparations. In Prague, 80 tonnes is consumed during the festive season. The trunks of gutted fish without heads or tails are sold for canning. Canned carp, often combined with vegetables, is a traditional food in European and Middle Eastern markets. Smoked carp paté by Glen Hill won a gold medal at the 2024 Melbourne Royal Australian Food Awards (Horn 2025).

But carp is not just a food. Parts of the fish can be used for a variety of purposes and products (https://www.kcfisheriesglobal.com/). For example, carp leather is made from cured and tanned fish skins. It is thicker than the skin of most fish and as strong as sheepskin or cowhide. It is used to make jewellery, phone cases, belts, wallets and bags, artwork, and binding for books. Carp scales are used in hobby markets for decoration, and in the paint industry for efflorescence. Carp collagen is used in health and beauty supplements. Carp pituitary glands are used in aquaculture as a hormone to stimulate fish to spawn. Carp swim bladders are used as finings in the production of wine.

Probably a carp product of which most Australians are aware, and many have used, is the fertiliser Charlie Carp® (https://charliecarp.com.au/). Based in Deniliquin, the company processes over 300 tonnes of carp annually to produce fertiliser from pulped carp and rice fibre. The solids are pelletised and the liquid is bottled. Established in 1998 by locals from Hay and Deniliquin, the company is now run by Ron Kopanica and Bob McFarland. Charlie Carp® has won several environmental awards, including the Prime Minister's Environment Award in 2000 and the Banksia Environmental Award in 2004.

NATIONAL CARP CONTROL PROGRAM AND THE CYPRINID HERPESVIRUS

We need to declare from the outset that we have both written on the topic of the cyprinid herpesvirus 3, otherwise known as CyHV-3, koi herpesvirus or KHV. Although we try to walk a straight line, you should be aware that we both played minor parts in the discussion

In it for the long haul: Keith Bell and K & C Fisheries

Box Fig. 10.1. Keith Bell netting carp, Gippsland, in the 1980s. Source: Keith Bell.

Keith and Cate Bell started catching carp commercially before most people even knew that the carp tsunami had arrived. They began the enterprise now known as K & C Fisheries (now including 'Global') in Sale, Gippsland in 1984 (http://www.kcfisheries.com.au/) and it continues to this day. They saw an opportunity to exploit an abundant resource, when others saw carp as a wasted effort. They literally shifted gear from native fish, and have never looked back. They have had the longest involvement with carp fishing and its commercialisation of anyone in Australia. The company aims to make use of carp while also helping the environment and the community. It harvests, processes, conducts research, does consulting jobs for local, state and federal government agencies and many other things. It has traded nationally and internationally, with markets and factories in Australia, the US and Germany, and works in Malaysia. It has about an 80% share in the commercialisation of Australia's carp as food for human consumption, rock lobster bait, petfood and fertiliser.

But more than all of this, the company has been a constant participant in carp removal projects, collaborating with many natural resource managers and scientists over the years. Keith in particular is a highly accomplished citizen scientist, with an insatiable curiosity about carp and its biology that has not dimmed over the years. He is the undisputed grandfather of carp in Australia. He and Cate have always given back to the community through various outreach programs, have been constant sources of enthusiasm on carp issues and supported the careers of scientists, including this author, Katie Doyle.

of whether the virus should be released. Since the mid-2000s the Invasive Animals CRC has touted CyHV-3 as a possible biological control agent (Fulton 2006) and the option has had strong promoters and vehement opponents. It is a controversial subject now and is likely to remain so for some time to come.

It makes sense to know as much as we can about the virus, its effects and the arguments for and against its release.

CyHV-3 is a herpesvirus in the Family Alloherpesviridae (Bergmann *et al.* 2020). There are closely related viruses, such as carp pox and goldfish haematopoietic necrosis virus. The distribution of CyHV-3 covers much of the world, except – perhaps – Australia and Antarctica, largely because of the global trade in common carp and koi. There are European and Asian forms of the virus, and although they can be told apart they are virtually identical. The virus probably originated in warmwater regions of Asia. Like many viruses, there are many variants and it can adapt quickly to different conditions. CyHV-3 was first isolated in the US in 1998, following mass deaths of carp in the US and Israel (Pokorova *et al.* 2005). It was detected in common carp in the mid-2000s in Western Europe, South Africa, Israel, Indonesia and Japan, and was already likely in other regions of Asia at the time (Pokorova *et al.* 2005).

CyHV-3 is mostly spread from fish to fish, although it can survive in water for some time (Bergmann *et al.* 2020). Fish can become infected at very low concentrations of the virus. It spreads within the host via white blood cells, from the gills and gut to other organs, then is released via the gills, gut and skin. The development of the disease depends on the virulence and concentration of the virus, the environmental conditions, the season, the stress of the infected fish, the genetic lineage of the fish and which tissue the virus infects. After three to nine days from infection, the virus causes lethargy and/or erratic swimming behaviour, lack of appetite, an enormous increase in mucus production on the skin and gills, and necrosis of the skin. Finally, if the infection is severe enough, death results (Bergmann *et al.* 2020). Temperature plays a big role in the onset and severity of the disease, because replication of the virus is temperature-dependent (Pokorova *et al.* 2005). From 13°C to 28°C the virus can be detected in fish, but death has only been detected at temperatures of 18–28°C. Above 35°C the 'infectivity' of the virus is destroyed, it effectively becomes harmless and the infected fish is inoculated.

The highest mortalities of carp because of CyHV-3 outbreaks have been in Japan and Indonesia in farmed and wild carp, when densities of fish were high (McColl *et al.* 2014). In Lake Biwa, Japan, for instance, 70% of carp died initially in an outbreak of the virus, although deaths gradually petered out over time. In most cases, however, the evidence has been that the proportion of fish that die because of CyHV-3 outbreaks was much lower than the Lake Biwa example (Thresher *et al.* 2018). Ron Thresher and colleagues analysed 17 mass die-offs between 2004 and 2018 in North America. They found that the die-offs usually were brief, not repeated over subsequent years and wild populations were less affected than previous reports under aquaculture conditions. They concluded that 'the short and long-term effects of most die-offs appear to be slight' (Thresher *et al.* 2018, p. 1703).

As part of the Invasive Animals CRC, Ken McColl and colleagues conducted a series of experiments to examine the infectivity of CyHV-3 on a range of Australian native freshwater fishes, some amphibians, reptiles, the yabby (*Cherax destructor*), chickens and lab mice (McColl *et al.* 2017). There were no mortalities in many of the native fishes, whereas there were some mortalities in Murray cod (*Maccullochella peelii*), golden perch (*Macquaria ambigua*), common galaxias (*Galaxias maculatus*), crimson-spotted rainbowfish (*Melanotaenia duboulayi*), Australian smelt (*Retropinna semoni*), carp gudgeons (*Hypseleotris* spp.), olive perchlet (*Ambassis agassizii*) and yabbies, but at similar rates to unexposed laboratory controls. Only in sea mullet (*Mugil cephalus*), silver perch (*Bidyanus bidyanus*) and a tadpole (*Litoria peroni*) were mortalities higher in the exposed animals than in the controls. McColl and colleagues argued that the timing of the deaths of the last group was not consistent with CyHV-3 infection, and furthermore there was no evidence that the virus had infected the fish. A critique of McColl *et al.* (2017) by Keller Kopf and colleagues (one of whom was this author, PH) took issue with the study on several grounds: 1) the conditions in which the fish were kept was clearly inadequate because of the deaths of control fish; 2) some exposed species did suffer higher rates of death than controls, and while it is unlikely they died from the virus itself, the factors involved in their deaths remain unexplained; 3) the ability of native fish to transmit the virus was not tested; and 4) sub-lethal effects of the virus, such as on immune function, stress, growth and reproduction, were not tested (Kopf *et al.* 2019a).

Some scientists and natural resource managers argue that the virus creates minimal suffering, kills relatively quickly and can potentially knock off a sizeable proportion of carp populations (McColl *et al.* 2014; National Carp Control Plan 2022b). With complementary measures (e.g. daughterless carp gene technology) and an integrated pest management plan, this may provide our best opportunity for decimating carp populations (Brown and Gilligan 2014). It is a once-in-a-lifetime opportunity to help our aquatic biodiversity recover. Repeated release of stronger versions of the virus every few years may enhance the efficacy of the initial release. The Victorian Fisheries Authority is certainly in favour of releasing the virus, and is heavily promoting it through multiple avenues (https://www.abc.net.au/news/2025-01-26/electrofishing-herpesvirus-used-to-control-european-carp/104713366; https://vfa.vic.gov.au/recreational-fishing/featured/australias-carp-problem). Some politicians, like ex-Deputy Prime Minister Barnaby Joyce, have been great fans of the virus for years, although his depiction of the virus as a type of venereal disease is more farce than fact (https://www.youtube.com/watch?v=vOvzAv1QTmU). A spoof of Joyce's parliamentary rant, courtesy of the ABC *Insiders* program, is priceless (https://www.youtube.com/watch?v=qg_KQYwKgpE).

Opponents of the virus's release are less optimistic about the outcome and the amount of focus on, and funding of, this approach. There is an argument that the money could be better spent on other types of carp control and river management (Kopf *et al.* 2019a). One of the main reasons some oppose the release of the virus is that it is unlikely to kill wild carp in anywhere near the numbers that are killed in high-density aquaculture situations, and 'there is little evidence to suggest that repeated CyHV-3 outbreaks would recur at a magnitude to counter the reproductive potential of the surviving carp' (Becker *et al.* 2019, p. 25). It is estimated that, if the virus were a one-off release, populations may recover within as little as two years if conditions are wet, or six years under more stable conditions (Hopf *et al.* 2024). Another criticism is that resistance to the virus will develop relatively quickly, perhaps within as little as 10 years after its release (Samsing *et al.* 2021). In other words, carp may become resistant to the virus, and even if they don't, populations could recover quickly anyway because of carp's reproductive potential,.

To give credit where it is due, the National Carp Control Program has done a huge amount of work investigating the potential of the virus to kill large numbers of carp, how it will be released, public engagement, the effects on water quality of lots of dead carp, and the socio-economic effects of carp control using the virus (National Carp Control Plan 2022b). Mind you, it had A\$15.2 million in funding to do the work (https://www.agriculture.gov.au/about/news/national-carp-control-program). There are individual reports on all these aspects (https://www.agriculture.gov.au/biosecurity-trade/pests-diseases-weeds/pest-animals-and-weeds/carp-biological-control-plan/national-carp-control-plan). Will the virus be released? If it is, how effective will it be? All we can say is: watch this space.

*****

Everyone agrees, though, that the CyHV-3 virus is not a silver bullet. It won't eradicate carp from Australia. At best it will reduce carp populations to 20–60% of their current size, which would leave a lot of carp swimming around in our rivers, lakes and wetlands. So, what do we do with the carp that are here now or that may continue to swim long into the future? Well, there are people in the community who value carp, not for commercial reasons *per se*, but recreationally. In the next two chapters, we present what some consider positive aspects of carp in Australia: recreational fishing and the keeping of ornamental koi. Believe it or not, some Australians cannot get enough of carp.

11

Recreational fishing

*When Britain's largest mirror carp 'Two-Tone' died on 14th August 2010,
over 50 mourning anglers gathered lakeside to commemorate the fish's long and
celebrated life (Cherrell 2022).*

CARP FISHING HAS CAPTIVATED ANGLERS THROUGHOUT EUROPE AND ASIA
for a very long time. *The Compleat Angler*, first published in 1653, called the carp 'the queen of
rivers; a stately, good, and a very subtle fish' (Walton 1909, p. 154). Not long after, *The Whole
Art of Fishing* also gave royal status to the carp, but added (with some mixing of metaphors and
sexes) that it was also 'the Fresh-Water Fox' (Anon 1714, p. 40). The carp's wily ways, coupled
with its size and the thrill of the fight, compels anglers in the UK, continental Europe, the US
and increasingly, Australia, to return again and again to their favourite waters in the hope that
this time they will catch *the* big one (Ragnarsson-Stabo 2015; Woon 2024).

From a minnow of a pastime almost 400 years ago, there is now a thriving recreational fishery
for carp in many parts of the world. In 2015, 7.4 million angling days and close to £1 billion were
spent trying to catch carp in England alone (Environment Agency 2018). You can book a dedicated
carp fishing holiday and hire a fishing guide in England, Slovenia, Croatia, Hungary, Italy and
France, where you will fish in picturesque lakes (carpcircle.com). Lake Como is not just home to
George and Amal Clooney, it is also where some of the largest carp in Italy can be caught by the
canny angler who likes to combine history, culture and carp (https://nice-baits.com/the-top-10-
most-beautiful-carp-fishing-spots-in-the-world/). Scores of books have been written on angling
for carp (Carp books online 2024) and carp fishing magazines – *CARPology Magazine*, *Total Carp*,
Big Carp Magazine, *Carp-Talk*, *North American Carp Angler*, *Crafty Carper*, *Advanced CarpFishing*
and *Carpworld* – adorn the shelves of newsagents in the UK and US. *CARPology* tells us that carp
fishing, apart from encouraging non-conformity and self-expression, is an extreme sport.

In England, particularly special carp are celebrated. Angling enthusiast Lee Jackson
chronicled his obsession with 'Two-Tone', an iconic fish among carp anglers, in *Just for the
Record: The Quest for Two-Tone* (Jackson 2013). 'For anyone who pursued Two-Tone, it became
an obsession for them and for those lucky enough to catch him it was the best day of their
angling career' (Jackson 2013).

And when Two-Tone died, his death was mourned.

Two-Tone: Funeral for a fish

When Britain's largest mirror carp 'Two-Tone' died on 14th August 2010, over 50 mourning anglers gathered lakeside to commemorate the fish's long and celebrated life. Many of the grieving fishermen brought flowers to the lake at Conningbrook in Ashford, Kent where a memorial service took place and a small plaque was unveiled. He was huge, powerful, handsome and a legend of the scene. When Two-Tone was last caught, the 45 year old fish weighed an impressive 67 lb 14 oz [30.7 kg]. He was known as a 'marriage wrecker'. People would dedicate every spare moment of their lives trying to catch the beast, neglecting their home lives (Cherrell 2022).

It is not just in the UK where carp are prized by anglers (Arlinghaus and Mehner 2003). Carp fishing worldwide is a billion-dollar industry, with tens of thousands of associated jobs and a devoted – some say, obsessed – following (Ragnarsson-Stabo 2015). It was once a male-dominated pastime, but its appeal has broadened in the last few decades. Indeed, the World Carp Classic – held each year at Lac de Madine, France – was won in 2013 by an all-female team. Carp fishing throughout much of the world, like much fishing, is considered more than just a hobby. It's an occupation. An addiction. Even, as Kate Cherrell says, 'a marriage wrecker' (Cherrell 2022).

Recreational fishing in China has been pursued for more than 2000 years, but since the early years of this century has exploded in popularity. As of 2018, 220 million fishers dangle a line annually (Arlinghaus *et al.* 2021; Barriault *et al.* 2021). The new recreational fisher in China seeks leisure, cultural heritage, scientific understanding and gastronomy all rolled into one. Much of the inland recreational fishing in China takes place in privately owned ponds, and carp is a prominent and prized angling species (Rajeshkumar *et al.* 2017). Similarly, recreational fishing is huge in Japan, high on the list of outdoor activities, especially for men (Arlinghaus *et al.* 2021). Carp is prominent among both river and lake anglers, even though since the 1970s there has been a general decline in catches (Katano *et al.* 2015).

The attitude toward carp and carp fishing in Australia, at least for most anglers, diverges significantly from that of recreational anglers in most other regions of the world. As one person put it: 'For Australians, catching carp is like catching leprosy' (Northern Suburbs Fly Fishing Club 2024). Considering the history of the introduction of the carp to Australia, its spread and its impacts, perhaps it is not surprising that this fish evokes such conflicting emotions. To understand the apparent contradictions that embody carp fishing, we first delve into the history of the internationally acclaimed carp angling scene, focusing on the UK where it seems to have taken a bizarre and extreme turn. Its culture of carp reverence influenced fishing in Australia, at least initially. Then we examine the history of carp fishing in Australia and the nature and motives of carp fishing competitions. Finally, we describe the rise in popularity of carp fishing in Australia and outline the tactics used to catch this fish.

CARP FISHING: UK VERSUS AUSTRALIA

As we have seen in Chapter 2, fishing for carp in the UK began sometime probably in the late 15th century as a means of getting food from a pond stocked for that purpose (Currie 1991). By the start of the 17th century, carp had expanded from a predominantly pond existence to live in rivers and it was by then also considered a prized angling fish. By the 18th century, land enclosures, industrialisation and urban expansion led to a division in angling (Locker 2014). 'Game' fishing, typically fly fishing, was traditionally an upper-class pursuit focused on salmonids (salmon, trout and grayling), carried out on privately owned or remote waters accessible only to a privileged few. 'Coarse' fishing included all other species – cyprinids especially – using baited hooks, generally in public areas that cost nothing to access. With industrialisation advancing and people moving to urban areas, coarse fishing became popular among the working class who fished along newly developed canals, ponds and rivers where water was not privately owned. These waters were commonly populated by cyprinid species, particularly roach and dace, which were more tolerant of the low oxygen levels and pollution characteristic of these environments. Cheap 'roach poles' were typically used (Locker 2014).

By the 19th century, coarse angling clubs had multiplied and prizes in competitions were awarded for the most fish or the greatest total weight (Locker 2014). By 1900, prize-focused 'match days' were increasingly popular (Lowerson 1983). 'Match fishermen' were a specific group of anglers focused on competition, and the events were becoming a central part of industrial working life. By the 20th century, there were signs of a shift in coarse angling preferences from catching large numbers of fish to pursuing record-sized specimens.

The introduction of selectively bred mirror carp from Holland and Germany in the late 19th and early 20th centuries (Clifford 1992) shifted the focus of carp fishing from food to sport. These 'king' carp tend to put on more body weight, have a deeper body and sustain bone growth for longer than the original wild scaled carp of eastern Europe (Balon 1995b). Mirror carp often exceeded 10 lb (5 kg) and were much larger than the commonly caught roach and European perch. Furthermore, they were bred to elude capture and became more evasive with age, so their angling status rose rapidly (Beukma 1970; Woon 2024). This shift sparked an interest in record-breaking catches, the establishment of fish icons such as Two-Tone, Mary, Clarissa and Fat Lady (Carpology 2020), and the emergence of celebrated UK fisheries like Redmires and Frensham (Woon 2024). Fat Lady was 61 lb (27.7 kg) and aged about 30 when she died of natural causes in July 2011. She had been caught and photographed at least 200 times. She was celebrated by a formal burial and a stretch of water was named 'Fatty's Point' in her honour at her home waters in St Ives (Derbyshire 2011).

The pursuit of record-breaking fish became a complex operation, involving lengthy preparations and specialised tackle. Initially not a favoured species (Hickley 1996), carp gained prominence by the late 20th century and it eventually transformed coarse angling from a leisurely pursuit into a highly competitive and profitable industry. The cashed-up angler

Table 11.1. Environmental, cultural and regulatory aspects, type of angling and genetic variations of carp fishing in UK and Australia.

	UK	Australia
Environment	Centuries-old traditions. Landscape of rivers, canals, ponds and lakes, offering diverse settings for anglers. Whatever impacts carp had on the environment began >400 years ago. Focus is on positive aspects of carp.	Relatively recent introduction. Landscape of rivers, billabongs and relatively recently constructed channels, weirs and lakes. Negative impacts of carp front-and-centre in people's consciousness. Focus is on negative aspects of carp.
Culture	Evolved into a celebrated sport with a passionate angling community. A blend of tradition, technique refinement and reverence for certain historic fishing locations. Anglers often engage in a quest for record-breaking catches and appreciate the nuances of different carp varieties.	A more utilitarian and conservation-oriented approach. A greater emphasis on controlling carp populations. Less about angling as a sport or hobby and more about environmental management.
Regulations	Operates within a well-established framework of regulations, licences and angling clubs. A culture of respecting fishing rights. Anglers often adhere to specific rules and etiquettes associated with different fishing locations.	Fishing regulations more centred on controlling and decreasing carp populations.
Techniques and tactics	Fishing often involves sophisticated tactics, specialised equipment and a diversity of baits and rigs. Anglers focus on understanding carp behaviour, employing methods like stalking, baiting and using 'boilies' to entice elusive fish. Culture of catch-and-release has influenced much of the angling gear.	More focus on control methods than on sport fishing. Involves techniques specifically designed to reduce carp populations, such as targeted removal programs including fishing competitions. Much of the fishing gear is adapted from other target species.
Genetic strains	Mirror and leather carp dominate. 'Wild' form less common in specimen fishing.	'Wild' scaled form most common. Mirror carp occasionally present but not common.

could buy tackle and accessories to increase the odds of a record catch while fishing in comfort (Carpology 2020). The pursuit and repeated capture of progressively heavier and older carp cultivated a catch-and-release culture, a heightened concern for the welfare of fish and an appreciation for the environment (Carpology 2017). In turn, this drove advances in the design of tackle and handling equipment (e.g. antiseptic liquid, carp sacks, weigh slings, unhooking mats, landing nets) that minimise damage to the fish (American Carp Society 2020).

Angling pioneers, like Richard Walker, influenced carp fishing by developing specific tactics and techniques. Walker's record catch of a 44 lb (20 kg) carp in Herefordshire in 1952 using a primitive fixed-spool reel and homemade tackle marked a turning point. 'Clarissa' was exhibited at London Zoo, sparking significant interest in carp fishing (Locker 2014). Even 70 years later, Clarissa's preserved body was listed as Lot 599 in a Mullock Jones auction on 24 July 2024 and expected to fetch up to £40,000, with the event featured in UK newspapers and available to view on YouTube (https://www.youtube.com/watch?v=XEaY08vQlas).

Walker and his associates went on to form the Carp Catchers Club, one of the most influential angling organisations in British history. The club shared information on baits, tackle and carp habits, and laid the foundations for the booming carp literature and multimedia culture that exist today (Carpology 2020; Carpworld 2023; The Carp Society 2024).

A typical English carpman spends 1000–2000 hours fishing each year, and considers it a good season if they manage to hook and land a dozen fish (Buffler and Dickson 2009). Yet in Australia, the humble carp is often overlooked by anglers; if it is acknowledged, it is commonly viewed unfavourably. Why do many Australian anglers look down on the fish that is esteemed throughout much of the rest of the world? The answer, in our view, can be found in the history of the dominance and the real and presumed impacts of carp on native fish and inland freshwater ecosystems more generally. The attitude began with the declaration of carp as a noxious species in 1962, the subsequent poisoning campaign and the relentless spread and successful colonisation of the Boolarra carp throughout south-eastern Australia and beyond.

So, the recreational carp fishing histories in Australia and the UK have been shaped by history, specifically the history of environmental change, culture and fisheries regulation (Table 11.1). Whereas the relationship with carp in the UK and continental Europe is founded on centuries-old traditions, Australia's carp fishing has been influenced by the environmental challenges of carp as a relatively recent invader. As we saw in Chapter 10, much focus has been on carp control and management (and occasional hopes of eradication, Chapter 9). Thinking of carp angling as a sport has come a very distant second. Nonetheless, the techniques (bait, lure and fly fishing, bow hunting) that have been developed and adopted in places like Europe and North America do exist in Australia, albeit to a lesser extent.

CARP COMPETITIONS DRIVE RECREATIONAL FISHING

In Australia, drivers of carp fishing (or lack of) can be broadly classified as: 1) an entirely negative perception of carp, meaning a strong dislike, and often a refusal to target carp; 2) fishing for food, most commonly by recent immigrants from Europe and Asia; 3) recreational fishing as a carp control strategy; or 4) targeted angling by those with a true passion for the species. The last group is the minority, but interest in the species as a target is slowly gaining traction. Most carp fishing in Australia is driven by two types of fishing competitions: carp control events or organised coarse fishing events.

Carp control competitions

The most widely embraced approach to carp fishing in Australia involves organised 'catch-a-carp' events, which promote the enjoyment of carp fishing while encouraging it as a way to remove fish and control carp populations. These began as early as 1978 in Lake Burley

Griffin, Canberra. Fishing competitions now occur across most of the distribution of carp in Australia. There are competitions in Victoria (e.g. Koroit Carp Fishing Competition (CoastFM 2023), [Fig. 11.1]), Queensland (e.g. Logan-Albert Catchment (BushnBeachFishing 2023)), Australian Capital Territory (Bredbo Australia Day carp out competition (Visit Canberra 2024)), New South Wales (e.g. Wentworth Catch-a-Carp (OzFish 2023)) and South Australia (Webster 2021). State fisheries agencies have developed guidelines for organising and running community carp fishing events (Norris 2011).

In New South Wales, competitions based in Sydney, such as Campbelltown Council's annual Catch a Carp (The Daily Telegraph 2017), get good coverage in Sydney newspapers. Another event, almost 13 hours' drive west to the small regional town of Menindee on the

Fig. 11.1. Carp fishing competition advertisement, Koroit, Victoria. Source: Koroit & District Angling Club Facebook.

Lower Darling-Baaka, is the OzFish Catch-a-Carp-a-Thon Menindee (OzFish 2023). One of Australia's largest competitions is the annual South Australian Carp Frenzy (Lake Bonney, Barmera), which in recent years has seen over 800 anglers attend and 16,000 carp pulled from the water in a single day (Dayman 2017; Webster 2021). There's even a Carp Tossing Championship at this event. In 2024 around $10,000 in prizes was offered – as if fishing for and landing carp was not enough incentive.

Prizes of cash, fishing gear, kayaks, fishing charters and weekend getaways are awarded for the heaviest, longest or 'mystery length' carp caught. Competitions, particularly in regional communities, attract non-locals who enjoy the opportunity to fish in remote locations. For locals, competitions are great money-spinners: out-of-towners spend up big on accommodation, food, tackle and fuel.

It is questionable whether these events have an impact on carp populations (Brown and Gilligan 2014), but that is rarely the sole purpose. Rather, they provide opportunities to teach kids and beginners to fish (North Central CMA 2019) and they educate people about the presence, roles and ecology of invasive and native aquatic species. And they are useful in making fishers aware of fisheries regulations more broadly. Carp are typically not eaten or released; instead, they are donated to fertiliser companies (e.g. Charlie Carp®, NutriSoil®). Carp fishing events provide recreational opportunities for all ages while generating funds for environmental initiatives, including native fish conservation and habitat restoration projects. For example, money raised from the Lake Bonney competition has helped fund the creation of 90 native fish hotels in Lake Bonney, supplementing the $20,000 from Recfish SA (Webster 2021). This trend is also evident in the US, in events like the Carp Slam competition in Denver (https://carpslam.org), which contributes funds to river health and habitat restoration efforts.

Bank angling competitions

Another type of fishing competition that is gaining popularity in Australia is 'bank angling', a type of coarse fishing (Fig. 11.2). In contrast to the carp removal competitions, proponents of this type of fishing promote catch-and-release and good management of the carp recreational fishery. There are coarse fishing and bank angling clubs in New South Wales (Sydney Coarse Angling Club, founded in the late 1980s), the Australian Capital Territory (Canberra Carp and Co.), Victoria (Melbourne Coarse Anglers, founded 1986), Western Australia (West Australian Coarse Anglers Club) and South Australia (South Adelaide Match Angling Group). The nature of coarse angling comes in several flavours: 1) pleasure angling, where anglers enjoy a relaxing day's fishing and are content to catch whatever fish they can; 2) match angling, during which anglers fish in teams or as individual entrants and gather at a location to catch either as many fish as possible in an allotted period, or the greatest total weight of fish (held at local, regional, national and international levels); and 3) specimen angling, the aim of which is to catch a large fish of a particular species.

At the international level, the Bank Angler Trans-Tasman competition (BATT) involves carp, conventional and coarse anglers from Australia and New Zealand. Anglers compete to see which team can catch the greatest total weight of carp over 96 hours (Bank Angler 2025).

Fig. 11.2. Active competitor in a bank angling competition. Source: Fidelis Qacha/Black Girl Gone Fishing NPC.

The competition attracts recent immigrants and Australian- and New Zealand-born anglers. To ensure fair play and environmental preservation, the BATT complies with state fisheries regulations, and each entrant must have a valid angling licence. Boats, kayaks, remote devices or fish-finding equipment are not allowed. Each team stays within designated peg-marked areas, and must respect the environment. Over the years, 5–8 kg carp have been commonly caught. The current record is 12.95 kg, set by Stuart Hammond in 2021. Members of bank angler clubs, which exist in most Australian states and territories, compete for selection in the national team that represents Australia at the annual World Carp Classic. Thousands of anglers around the world compete every year for the title of Carp Fishing World Champion.

Importantly, Australian bank angling competitions are not carp control events. A philosophy of coarse anglers is to practise catch-and-release of live fish. The competitions operate within the established international ethics of carp angling as a sport. Whenever possible, fish are returned to the water alive. In New South Wales the release of live fish is permitted for bank angling competitions, and is sometimes also allowed under a special permit issued by other state fisheries authorities (e.g. in Victoria, participants in the Australian Federation of Coarse Anglers Association events are permitted to hold live carp and return them within 10 m of the point of capture). For coarse anglers, fishing is a sport in which respect for the fish, and maintaining fish welfare, are paramount. Competitors must possess fish-friendly carp sacks or knotless keepnets to ensure the safe handling and return of recorded fish to the water. All fish are released and swim off under the supervision of weigh-in officials.

The Australian Federation of Coarse Anglers Association, which represents coarse anglers and coarse angling within Australia, was founded in 1986. It sees carp as an unexploited resource, and believes that the UK industry of recreational carp fishing could be mirrored in Australia:

> *To showcase the sport to the Australian public to increase participation numbers.*
> *Also 'exposing them to our passion and carp's untold story will also help people*
> *to have a better understanding of why we promote a more sensible approach*
> *to carp control by properly managing the species' internationally renowned*
> *potential as a sport fish and resource'. Country towns will reap the rewards*
> *whilst whole communities as well as anglers and their families thoroughly enjoy*
> *the social benefits associated with the sport. It happens all over the world and*
> *there is no reason why it can't be the case here in Australia and New Zealand*
> *(Bank Angler 2024).*

Is carp destroying or creating a fishery?

In Australia, anglers come at carp fishing from two vastly different perspectives, embodied in the different types of fishing competitions. One side holds the view that carp is responsible for destroying our native or game (mainly trout and redfin) fisheries, and advocates for its removal, with fishing as a potential solution. This perspective generally aligns with initiatives that label carp as vermin that need exterminating. The belief is that carp is detrimental to recreational fisheries, not that it contributes to the development of a new fishery.

On the other hand, coarse fishers consider that carp *is* the fishery. They promote catch-and-release practices as a sustainable fisheries management approach. Catch-and-release fishing for carp is controversial (Fig. 11.3), as other anglers believe the survival of native fish, particularly other recreationally significant fish such as Murray cod and golden perch, are at risk through the practice of returning carp. At times, disagreements can boil over:

> *Recent reports of the Sydney Coarse Angling Club's annual Canberra Classic*
> *angling competition in Lake Burley Griffin attracted a number of ill-informed and*
> *inappropriate responses, including an assertion that club members who released*
> *captured carp back into the lake were 'environmental vandals'. One member*
> *of the club was even subjected to threats after the publication of two articles in*
> *this newspaper which sought to highlight a potential conflict between the ACT*
> *government's carp reduction strategy and the angling club's commitment to catch-*
> *and-release fishing. The club refutes the claim of vandalism – put simply, nothing*
> *was destroyed and releasing carp back into the water from which they were caught*
> *is not illegal in the ACT. The club has worked closely with ACT authorities since it*
> *first hosted the event here in 1995* (Sydney Morning Herald, *21 December 2012).*

Fig. 11.3. Anglers releasing carp after catching. Source: WikiCommons, CC BY-SA 4.0.

Concern has also been expressed about the deliberate transfer of carp to new waters to establish new coarse fisheries. Advocates of catch-and-release of carp value the cultural significance of carp fishing, and aim to marry the enjoyment of the sport with the preservation of native species and their habitats (Colvin *et al.* 2012). They generally oppose the koi herpesvirus release and prefer to utilise and manage carp as an angling resource (Bank Angler 2024).

Are bank anglers any different from those who pursue angling for brown and rainbow trout as a pastime?

TYPES OF CARP FISHING

Fishing for carp involves several methods, broadly grouped as bait fishing, artificial lure fishing, flyfishing and bow fishing. We describe each below.

Bait fishing

In broad terms, there are two main types of bait fishing for carp: 'rough' style fishing, using a simple rod and reel set-up and is relatively low cost, and often gear from other fishing methods can be used; and coarse fishing, a highly technical and specialised method originating from the UK. Both methods are largely undertaken on the bank of a river or lake; hence the terms 'bank fishing' or 'bank angling' using baited hooks (see Table 11.2 for a summary of the different types of baits). Typically, two rods are stationary on rests, allowing anglers to engage in other outdoor activities such as cooking, family time and relaxing, while waiting for a bite (Bank Angler 2024).

Table 11.2. Types of baits used for carp fishing in Australia and globally.

Bait type	Description
Boilie	A mix of ingredients such as fishmeal, dough balls, birdseed, grains, milk proteins, eggs and artificial flavours (pineapple and strawberry are popular in the US) is boiled in water to make the bait more robust. Boilies come in various sizes, colours (brightly coloured or more natural in appearance) and flavours, and are a staple in carp fishing. Baiting is often performed using baiting tools to increase accuracy and range. For bigger fish and longer fishing sessions, boilies are excellent to use on a hair rig. Recipe books allow anglers to make their own boilies (Homemade Boilies, UK 2023).
Pellets	Pellets are compressed baits in different sizes made from ingredients like fishmeal, halibut or maize. They can be used as hook bait or to create a feeding area.
Seed particles	Natural baits such as maize, hempseed, chickpeas or tiger nuts, these are commonly used as a hook or hair rig bait. They can be prepared, cooked or flavoured to attract carp.
Sweetcorn	A popular and cost-effective bait. Carp are attracted to its bright colour and sweet taste.
Bread	Bread can be flaked, mashed or balled to suit different fishing situations. Pieces of bread or dough can be effective when fishing on the surface.
Meat	Luncheon meat or sausages. These can be flavoured or prepared to enhance their attractiveness.
Natural baits	Insects like maggots, worms (e.g. nightcrawler earthworms) or caterpillars.
Artificial baits	Synthetic baits, such as imitation particles (corn, maize, tigernuts, breadflakes, pellets), plastic boilies, artificial maggots and even luncheon meat versions have become popular over the last 15 years. They offer several advantages over 'real' baits, including resistance to attack by crayfish or unwanted fish species, and can absorb different flavours to increase attraction. Some imitation baits include fluorescent colours or niteglow materials to visually attract carp (Mitchel 2019).

The popular carp control competitions and family gatherings along a river exemplify the simplicity of carp (or 'rough') fishing. This is the most common type of carp fishing in Australia. Baits, such as corn, worms and bread, are cast out on a light rod.

Rough fishing is characterised by its simplicity:

> *... rough fishing also offers anglers a chance to return to the basics of fishing. By wading rivers for carp ... you can walk away from the philosophy of electronic gadgetry that permeates the fishing information anglers receive from TV and magazines ... If nothing else, rough fishing lets you re-examine your sport from another angle, one that doesn't depend on boats, motors, digital reels, and color-coded lures for catching fish and having a good time, but instead makes use of your common sense and appreciation for the outdoors (Buffler and Dickson 2009, p. 5).*

In contrast to rough fishing, coarse fishing (sometimes called 'Euro style' as it originates primarily from England, the Netherlands and France) has developed over centuries to achieve technical accuracy and specialisation of equipment, often with a focus on catching fish repeatedly. Bank Angler (https://www.bankangler.com.au), Australia's primary website dedicated to carp angling styles using naturally baited hooks, warns that this style of fishing can be overwhelming. The huge range of flavours, hook baits, terminal tackle, rods, reels, tackle boxes, tackle box stands, rod rests and keepnets, associated camping equipment and other

Fig. 11.4. Top left: Fishing rod pod and electronic bite alarms used by carp anglers. Source: Wikimedia Commons, CC BY-SA 4.0. Middle: Carp-specific rigged bait. Source: ZAS Backhouse. Bottom left: Carp fishing camp with rods. Source: © Copyright Bob Abell and licensed for reuse under CC BY-SA 2.0. Right: Two rods on a double tripod stand equipped with tackle box, nets and bait buckets. Source: S.C. Van Niekerk).

equipment, all specialised for catching carp, is astonishing (also see https://www.carpangler. com/ for the sheer amount of available equipment) (Fig. 11.4).

Coarse fishing can be categorised into three main groups: bottom fishing (or ledgering), float fishing and surface fishing, based on the part of the water column the angler targets. Bottom fishing is the most popular and it aligns with the bottom-feeding behaviour typical of carp. Bottom fishing involves presenting bait on the lakebed or riverbed using weights (or method feeders/PVA bags) to keep it stationary at depth.

Float fishing, one of the oldest and most traditional carp fishing methods, employs a float (e.g. waggler float) to present a bait that is suspended at a desired depth. Groundbait or small particles like hemp or birdseed can be used as a chum to create more attraction around the hook bait. A common strategy is baiting ('chumming') various locations before fishing, and revisiting them later to observe signs of feeding fish.

Surface fishing for carp involves targeting the fish as they actively feed on the water's surface. Surface fishing uses floating baits like bread packed around the hook, and other buoyant baits like sunflower seeds, feed in a bag (PVA bags), dog biscuits or artificial baits such as fake bread or shaped cork (AnglingTimes 2018). As with float fishing, chumming areas with buoyant baits is done to encourage carp to feed.

Artificial lure fishing

The use of artificial lures is a common fishing method for carp in Australia. Typically, lures are cast accurately at fish that the angler has seen in the shallow margins of streams, rivers and lakes. Various types of artificial lures are employed for carp fishing, the most popular of which are soft plastic crustaceans, minnow soft plastics, grub soft plastics, paddle tail soft plastics and small crankbaits/hardbody lures. This method often utilises light spin fishing gear intended for other target species. Carp is frequently caught as by-catch when fishing for other freshwater species, most commonly Murray cod, golden perch, redfin and trout.

Fly fishing

Fly fishing originated in the UK, predominantly for game species like salmonids (Whitelaw 2015). Fly fishing for carp is a popular pursuit, attracting enthusiasts and even becoming a tourist attraction, as evidenced by the Portuguese holiday experience offered at https:// carponflyadventures.com/. The trend is observed in Europe and North America as well. The American Carp Society has a dedicated website (https://americancarpsociety.com/guiding) of 'officially sanctioned, experienced carp fishing guides'. The most successful way to target carp while fly fishing is to cast at sighted fish. Fly fishing for carp involves using specialised flies designed to imitate, in appearance and behaviour, insects or other natural food sources that carp commonly feed on. Fly patterns specifically for carp include: imitations of aquatic bugs and crustaceans such as crayfish/shrimp; nymph or worm patterns such as Woolly Buggers, Woolly Worms, San Juan worms, Hybrid (Fig. 11.5) and Carp Crack flies; imitations of terrestrial beetles, cicadas and grasshopper flies for surface-feeding fish (Cameron McGregor, pers. comm.); plant

Fig. 11.5. An adapted US fly pattern 'John Montana's Hybrid' carp fly used in Australia, Lake Hume, New South Wales. Source: Katie Doyle.

material (mulberries or cottonseed); and introduced foods (imitation bread, dog food, corn). Most are influenced by US fly patterns (Elkins 2024). Dry flies for carp on the surface can be effective where carp is actively feeding on an insect hatch or seasonal crop of mulberries or cottonseed (Anon 2017). As the Australian carp fly fishery evolves, more fly patterns are being adapted or created for the local fishery.

Carp is probably one of the few fish species that has alternative names in the recreational fishing scene. This is most evident in fly fishing articles and conversations (Frasier 2023). It is known as freshwater redfish (Anon 2017), golden bones (Tucker 2016), the golden ghost (Whitlock 2021) or freshwater bones (named because of its similarity to bonefish in size and fight), in allusion to the world's best fishing destinations in the Caribbean or Florida Keys. Other names, commonly linked to urban carp populations, are mud-marlin, sewer salmon, pond pig, river rabbit, dumpster dolphin and ghetto grouper (Pearson 2019).

Bow fishing

Bow fishing is an angling technique that uses specialised archery gear, comprising an upright bow with an arrow attached to a line and reel for shooting and retrieving fish (Fig. 11.6). It is particularly effective for larger surface-dwelling or shallow-water species like carp, as it allows anglers to visually stalk and harvest fish on foot or from a boat. Originating from the US, where it is popular as an outdoor sport and is legal across most of the country, bow fishing has recently become legal in some parts of Australia.

The legality of bow fishing varies across Australia, with Victoria still classifying it as illegal (Victorian Fisheries Authority 2022). In South Australia, bows and arrows can be used during daylight, in the Murray River outside of the main stem and at least 50 m from anyone not fishing (PIRSA 2022). In New South Wales, bow fishing is permitted with a recreational fishing licence

Fig. 11.6. A bowfisher takes aim at a carp. Source: Yoyo500, Wikimedia Commons, CC BY-SA 4.0.

when targeting carp in most inland waterways (NSW Department of Primary Industries 2024), subject to rules regarding equipment, distance limits, time of day restrictions and humane fish dispatch and disposal (Graham *et al.* 2021). Since its legalisation in New South Wales and South Australia bow fishing has seen a surge in popularity, as evidenced by the presence of instructional videos and bow fishing stories on platforms like YouTube showcasing activities post-2020.

Bow fishing in Australia has been opposed by some academics, environmental groups and politicians who have concerns about unintended harm caused to native and threatened species due to the difficulty of distinguishing them from carp in turbid waters (O'Brien and Cormack 2016). Bow fishing makes catch-and-release impractical, and wounding mortality or injury can occur to uncaught fish (Montague *et al.* 2023). Additionally, the selective removal of larger, older fish can lead to evolutionary disruptive truncation of life histories (Scarnecchia and Schooley 2020). Critics often doubt its efficacy in reducing carp populations and question bow fishing's potential impact on vulnerable species, like rakali (native water rats, *Hydromys chrysogaster*) or platypus (*Ornithorhynchus anatinus*) (Ferguson 2016).

Modern urban fishing

Finally, while the draw of record catches from world-famous picture-postcard European lakes is attractive, there is a resurgence in a type of coarse fishing in the UK: urban fishing. The urban fishing scene in Australia and North America is also gaining traction through online and print articles about the best locations to catch fish (Lapin 2020; Pavia 2023), including koi with orange or black colourations (https://americancarpsociety.com/koi-carp).

*****

Despite historical neglect or negative attitudes in certain angling communities, carp angling has become one of the most prominent recreational freshwater fisheries globally, primarily due to the challenge it offers. In Australia, carp is slowly gaining recognition for its fighting abilities and is attracting a new following. The abundance of carp in Australian waters and shifting attitudes have led to more anglers taking up the sport as an alternative to fishing for other, more popular species. There remains tension between those who view fishing for carp as a means of reducing populations and want to see it eradicated from Australian waters, and those who value carp for its angling qualities and release fish once caught. But carp in Australia is valued by some not only for angling , but for its beauty. And it is to the aesthetic aspect of carp that we turn in the next (and penultimate) chapter.

12

Carp as ornamental fish

Your good friend doesn't deserve to be dumped (NSW Government educational material for rehoming unwanted fish).

OUR PRIMARY FOCUS IN *CARP IN AUSTRALIA* HAS BEEN ON THE COMMON-OR-garden drab, brownish-grey-green wild common carp, found throughout many of Australia's waterways. However, no discussion of carp is complete without consideration of its ornamental counterpart, the koi, or *nishikigoi* in Japanese. 'Koi', meaning 'carp' in Japanese, is a variety of common carp which is enormously popular globally. Koi keeping, a centuries-old Japanese phenomenon, has relatively recently taken off in Australia. Koi ponds are now a prized feature in some high-end Australian homes.

Koi is the real-life equivalent of the hero of Hans Christian Andersen's story, transforming over many years of selective breeding from 'ugly duckling' to 'beautiful swan' of the fish world (Tamadachi 1994). Unlike its wild relatives, koi comes in an array of colours – white, red, black, blue, gold, yellow, silver, green and even metallic – as well as unique scale patterns and body shapes. Today, koi is celebrated for its beauty and high value. It is the world's most recognisable and in-demand ornamental fish, with millions sold each year, some at exorbitant prices. Koi holds deep cultural and symbolic significance, especially in Asian cultures. The amount of breeding, research into genetics, exhibitions and decoration of ponds, temples and public spaces worldwide is mind-boggling.

How did this 'ugly duckling' transform into such a 'beautiful swan'? In this chapter, we explore how an ordinary fish, not known for its good looks, over the centuries turned into a celebrity in China and Japan. We also look at how the cult of the koi made its way to Australia. The paradox is that a fish that has been declared noxious in several Australian states and territories, often associated with turbid murky waters and widely dismissed by most Australians, is the same species as the one that enthusiasts keep in crystal-clear water to admire its beauty. Ardent admirers are even willing to pay top dollar for a special fish. We ask: how do these opposing attitudes coexist? Again, our intention is not to encourage the keeping of carp for ornamental reasons, we merely explain the phenomenon.

ORIGINS AND HISTORY

Beginnings

Koi can be traced back to two key periods: 1) prior to the 1800s, when ornamental fish domestication and colour appearance began; and 2) post-1800s, when selective breeding led to modern koi, the rise of koi-keeping as a hobby and the emergence of koi exhibitions. Here, we rely heavily on the research of Michugo Tamadachi, especially his *Cult of the Koi* (Tamadachi 1994), and of Eugene Balon (Balon 1995b), who we mentioned in earlier chapters.

As we have seen, there has been a long history in many parts of the world of fish being kept in ponds for food. When a species is kept in a captive breeding population, natural mutations can alter its appearance and colour. In the wild, such mutations may be a problem because an affected fish is more likely to attract the attention of a predator, but in captivity, visually appealing fish may be kept alive and bred by their owners (Balon 1995b). We do not know when exactly the first mutant coloured carp appeared, but this is not uncommon with most domesticated pet species (Tamadachi 1994).

The practice of raising fish for their coloured appearance likely began with goldfish in China, pre-dating the interest in coloured carp. From their original greyish-brown wild form, red-scaled goldfish were first recorded during the Jin Dynasty (265–420 CE), with breeding for ornamental ponds occurring by the Tang Dynasty (618–907 CE) (Komiyama *et al.* 2009; Chen *et al.* 2020). By the Song Dynasty (960–1279 CE), yellow (gold), orange, white and red-and-white varieties of goldfish were cultivated (Tamadachi 1994).

Balon provided two of the first significant scientific reviews of the origin and domestication of common and ornamental carp, using paleontological records, historical texts and morphological analysis (Balon 1995a, b). But even he was uncertain of the exact origins of coloured carp. While coloured carp may have existed for centuries, authenticated records before the 1800s are scarce (Tamadachi 1994). Ancient references to coloured carp often describe crucian carp or goldfish. In the absence of written evidence, artworks, including old silk paintings, pottery and carvings, are referenced, but these frequently depict wild common carp not coloured koi. Confusion between common carp and goldfish is frequent, with artists sometimes adding or omitting barbels – as we know, a key feature distinguishing the two species (Balon 1995b). This misidentification persists even today, as koi and goldfish are still often mistaken for one another.

Early colouration in carp likely resulted from natural mutations rather than from intentional breeding, creating a patchwork effect on predominantly brown scales (Balon 1995b). These changes probably went largely unrecorded. Colour mutations are common, but they were likely impractical for food farming in rural communities due to unstable inheritance and high costs of maintaining the traits. Balon speculated that the early mutations were red, and may have existed since Roman times. But again, there is confusion whether the Romans were referring to carp or goldfish (Balon 1995b).

Some believe that coloured carp may have originated in China (Axelrod 1973; Davies 1989; Tamadachi 1994). The Chinese are known to have domesticated other animals such as mice, dogs, birds and goldfish, and to have developed a golden variety of common carp (Tamadachi 1994). Japanese carp authorities Tamaki and Aoki (Tamaki and Aoki 1977;

Tamadachi 1994) reference records dating to 200 CE that mention red, white and blue carp, and another record from around 700 CE that an emperor viewed and admired 'koi', with no further details. American koi expert Herbert R. Axelrod (Axelrod 1988) agreed with Japanese authorities cited in Balon (2006) that modern coloured koi did not emerge before the 1800s, but suggested that coloured carp and goldfish existed in China as early as the 1300s. Takeo Kuroki noted that 'wild koi' were kept by the nobility between 785 and 1192 CE, but it was the people of Niigata Prefecture in Japan who developed red and white koi colours in the 1800s (De Kock and Gomelsky 2015). Advances in genetic research, such as mitochondrial DNA analysis (Mabuchi and Song 2014) may eventually clarify the precise historical timeline of the coloured carp.

Modern koi

Unlike the ambiguous origin of 'coloured carp,' the history of modern koi is well-documented. Japan's Niigata Prefecture is acknowledged as the birthplace of selective koi breeding. Between 1781 and 1788, rice farmers in the mountainous Yamakoshi region of Niigata Prefecture began raising young carp, sourced from nearby rivers, in terraced rice fields and wooden ponds as a protein source during winter (Koshida 1931; De Kock and Gomelsky 2015). The region's long winters, isolation and human curiosity are believed to have played significant roles in the development of coloured carp. Balon (2006) suggested that the fish held in prolonged darkness under snow may have influenced colour abnormalities through melatonin production, while De Kock and Gomelsky (2015) suggested that drought and famine may have limited the fish stock gene pool, thus increasing genetic mutations.

By the early 1800s, farmers noticed naturally occurring colour mutations from cross-breeding the original wild carp, *magoi* (black or true carp) (De Kock and Gomelsky 2015). They began cultivating the carp for visual appeal rather than food, separating the colourful ones into designated ponds (Tamadachi 1994). While there is no documented record of the first colour mutation, as with goldfish, it is believed that the first colour to appear was red (Tamadachi 1994; Balon 1995b).

By the 1820s, farmers had refined their breeding techniques to maintain the colour mutations and develop new colour forms, resulting in fish collectively termed *kawarigoi* ('changed carp') (Tamadachi 1994; De Kock and Gomelsky 2015). From the early 1800s to 1830, breeders developed red-and-white koi varieties through crossing red carp with white carp, including *hara-aka* ('red-belly'), *hoo-aka* ('red-cheeked'), *kuchibeni* ('red lips'), *zukinkaburi* ('some red on the head') and *menkaburi* ('all red on the head'). Later, *sarasa* ('red markings on the back') appeared – the precursor to the most prized *kohaku* ('white with red on the dorsal and lateral surfaces'). While the red-and-white varieties were particularly popular, cross-breeding with *magoi* (original black carp) also took place, resulting in *asagi* (red and white with blue along the back) and *bekko* varieties (*shiro bekko* – white and black, *aka bekko* – red and black, *ki bekko* – yellow with black), forming the basis of modern koi colours (Tamadachi 1994).

Although the farmers were unaware of genetics, their practices coincided with Gregor Mendel's work on heredity in pea plants (1856–1863), which formed the foundation of modern genetic studies. Today, koi breeding still relies on selective breeding knowledge handed down through generations (Tamadachi 1994). These fish, bred for their colourful scales rather than as food, are known as 'living or swimming jewels' or 'swimming flowers' (Amano 1968). New varieties impressed visitors to Ojiya City in Niigata Prefecture, sparking the growth of ornamental koi-keeping as a hobby.

During the 19th century, many koi colours and patterns became firmly established, reducing the reliance on wild-type carp for breeding. In the 1920s, Sawata Aoki and his son caught wild carp with golden scales, possibly descendants of escaped Nagasaki imports, and cross-bred them to create the golden *ogon* variety, which was first displayed in 1946 (De Kock and Gomelsky 2015). Around 1904, domesticated scaleless mirror carp imported from Germany were introduced into the breeding process, creating new scale patterns and the variety known as *doitsu nishikigoi*. In Japanese, *doitsu* means German, and today many koi varieties feature *doitsu* scales (Balon 1995b).

Hobby and exhibitions

The first koi exhibition is believed to have been held in Niigata in 1912, for professional breeders (Tamadachi 1994). In 1914, coloured carp was exhibited at the Tokyo Taisho Exposition for the first time. That year, Emperor Showa (aka Crown Prince Emperor Hirohito) received eight fish to be kept in a palace moat, which boosted the fish's popularity and market value across Japan. This renewed farmers' determination to continue koi breeding despite their isolation, harsh environmental and political challenges and limited access to resources and technology. In 1929, the first Yamakoshi Farming Koi Show was held in Higashiyama.

Between 1915 and 1942, new varieties emerged, including the *gosanke* group. This group features *kōhaku* – known as the 'King of the *Nishikigoi*', one of the first varieties established in Japan in the late 19th century. The group also includes the three-coloured varieties *taishō sanshoku* or *taishō sanke*, first exhibited in 1914 during the reign of the Taishō Emperor and established as a breed in 1917, and *shōwa sanshoku*, first exhibited in 1927 during the reign of the Showa Emperor and established as a breed by 1935. These varieties remain the most popular today (De Kock and Gomelsky 2015).

World War II led to food shortages, resulting in many carp being killed and eaten, but some were preserved in temple ponds and breeding knowledge was safeguarded by the farmers (Tamadachi 1994). After the war (post-1950), koi breeding resumed, aided by improved living standards in Japan and the West. Enhanced transportation and packing innovations (e.g. plastic bags and packing materials) facilitated fish exports, while pond liners and improved pond technology made pond ownership more accessible. Public zoos and aquariums and organised koi shows from the 1960s increased demand, transforming the once-exclusive fish into a widely accessible hobby. As koi started to spread globally, exhibitions were held in the US, UK, Europe and beyond.

Nishikigoi

Before the 1910s, coloured carp went by various Japanese names: collectively *kawarigoi* or *irogoi* (colour carp), or pattern-specific terms like *hanagoi* (flower carp), *moyogoi* and *madaragoi* (spotted carp) (De Kock and Gomelsky 2015). When koi were first exported to Western countries, the diverse English translations led to confusion in the terminology of koi varieties (Tamadachi 1994). The name *nishikigoi* meaning 'brocaded carp' was proposed by Kei Abe in 1918, who considered the *taishō sanshoku* variety a thing of beauty. *Nishiki* refers to a luxurious brocade fabric woven with red, white, black and metallic threads, traditionally used for high-class Japanese garments (obi and kimono) (Balon 1995b). *Goi* derives from *koi*, meaning carp. However, it wasn't until the late 1960s and early 1970s that *nishikigoi* became widely accepted, largely due to the efforts of enthusiasts like Hideo Miya and Takeo Kuroki.

Hideo Miya popularised the term by using it in the catalogue for the inaugural All Japan (Nippon) *Nishikigoi* Show, Tokyo in 1968, while Kuroki Takeo advocated for *nishikigoi* as 'a grand hobby and cultural pursuit' (Tamadachi 1994). Takeo established the Oita Airinkai Koi Club in 1962, which grew into a national organisation, the Zen Nippon Airinkai (ZNA) in 1968 and became the *Zen Nihon Nishikigoi Shinkokai* in 1970, renamed the International Nishikigoi Promotion Center in 2003 (De Kock and Gomelsky 2015). His global outreach and publications significantly advanced *nishikigoi*, laying the foundation for today's koi hobby.

For those interested in keeping koi, knowing the Japanese names of koi is not necessary – simply pointing out a favourite fish to a dealer is sufficient. Breeders, however, classify koi by colour, patterning and scalation. Single-coloured koi, such as white (*shiro-muji*), red (*aka-muji*) and yellow (*ki-goi*) are distinguished not only by their colour but also by whether they have a metallic appearance, with *ogon* indicating a metallic sheen. Koi with two colour patterns include red on white (*kohaku*), blue-grey with red along the sides (*asagi*), black on white (*shiro-bekko* where *bekko* means tortoise-shell), black on red (*aka-bekko*), white on black (*shiro-utsuri* where *utsuri* means reflection) and yellow on black (*ki-utsuri*). Three-coloured *sanshoku* koi include white koi with red (*hi*) and black (*sumi*) markings (*taishō sanshoku* or *taishō sanke*) and black koi with red (*hi*) and white marking (*shōwa sanshoku* or *sanke*), while *goshiki* means five colours. In addition to colours and patterns, there are naming variations for scale formation. Some koi have typical carp scales; others, like *doitsu-goi*, have rows of large scales or are scaleless (https://www.akakoi.com.au/about-koi).

New koi varieties such as butterfly koi, dwarf koi, ghost koi and solid white continue to be developed, several outside Japan (Bartz 2017; Van Zen 2023). Some consider these varieties to be outside the traditional definition of *nishikigoi*; as a result, they are generally not permitted in Japanese koi shows but are popular in countries such as the UK and the US (Lefever 1993; Balon 1995b).

With over 120 different *nishikigoi* varieties (Zen Nippon Airinkai 2024), koi organisations and suppliers often provide photographs, videos and lineage charts to explain the unique colour and pattern development of each fish (Fornaro 2020; Kodama Koi Farm 2024; Koi Society Australia 2024).

CULTURAL SIGNIFICANCE AND SYMBOLOGY

Koi (and carp) hold deep symbolism in Chinese and Japanese cultures, associated with positive qualities such as perseverance, strength and prosperity (Pattist 2015). In China, the symbolism stems from the Chinese legend of the Dragon Gate, in which a determined carp swims upstream, overcoming numerous obstacles and waterfalls to reach the Dragon Gate at the Yellow River. Upon reaching the gate, the gods transform the fish into a powerful golden dragon. This story has inspired Chinese art and literature as a representation of transformation and strength.

In Japanese culture, koi is associated with qualities like luck, prosperity and tenacity. It is the country's national fish (Damaschin 2021). The cultural prominence of modern koi began relatively recently, in the early 1800s, but most Japanese people saw koi first-hand only since the 1960s (Anon 2010). This rise is notable within a culture rooted in ancient traditions like *katana* (swords of the samurai), *hanabi* (fireworks) and *onsen* (baths).

In Japan, koi is presented as a gift as a symbol of love, gratitude and peace (McDonough 2023). The fish is also linked to samurai bravery, symbolising strength and perseverance as it swims upstream against a river's current. The phrase *koi no taki-nobori* means 'carp climbing the waterfalls', and is based on the Chinese symbolism for overcoming challenges and achieving goals (Damaschin 2021) (Fig. 12.1).

Fig. 12.1. Japanese cloth depicting 'carp climbing the waterfalls'. Source: Photographed by Katie Doyle, artist unknown.

The traditional custom of *koinobori* (carp streamers), which began in the Edo period (1603–1868), is still practised today and displayed in gardens on Children's Day (5 May) as a celebration of hope and resilience (Damaschin 2021). Koi also became an icon of recovery and resilience after the Niigata Chūetsu earthquake devastated the koi industry in 2004 (Damaschin 2021).

Koi has inspired artists worldwide, attracting audiences with its vibrant colours, graceful movements and blend of artistry and nature. Koi is featured in traditional Japanese art forms, including folk crafts (like *koinobori*) and kimono designs. Over time, artists began incorporating coloured carp and koi into their works. Koi is popular in contemporary art, bridging traditional and modern styles. It can be seen in traditional ink wash paintings as well as in modern digital art, in watercolour, acrylic and oil paintings, colouring-in books, origami, metal sculptures, stone carvings, jewellery and even coins (Fig. 12.2). Often, the art follows common themes of yin and yang, pond scenes with lilies, schools of koi swimming and fish struggling upstream against waterfalls. Koi is featured in art collections around the world, including those at the Royal Thai Art International Co. (Royal Thai Art International 2024), the Metropolitan Museum of Art (Anon 2017), Fine Art America (Fine Art America 2024) and the China Online Museum (China Online Museum 2024).

Koi-themed tattoos, or *irezumi*, are extremely popular, with fish colour and swimming direction carrying specific meanings (Fellman and Thomas 1987; Ashcraft and Benny 2016). Black symbolises masculinity, gold signifies prosperity and wellbeing, blue is often linked to the role of a son or represents tranquillity, and red embodies strength and power, also recognised as the matriarchal figure (Becker *et al.* 2019; Corby 2022). Koi tattoos are personal expressions of a person's life story and values.

For its vibrant colours and patterns, koi is celebrated as a living work of art. A notable koi display is the Monet Pond near Nemichi Shrine in Gifu Prefecture, which became famous in 2015 for its resemblance to a Claude Monet painting and earned the nickname 'the pond where

Fig. 12.2. Australian coin 'Chinese myths and legends: red dragon and koi 2023'. Source: Perth Mint, Australia.

art comes to life' (Patowary 2019). The pond is fed with nutrient-free spring water, and the resulting clarity gives a clear view of koi swimming beneath the lilies. Koi tourism is popular in Japan, and attracts people from around the world who want to explore the cultural and aesthetic value of these fish. *Nishikigoi* was officially recognised as Niigata's ornamental fish in 2017 (Niigata Prefecture Tourism Association 2024). Koi viewing is a popular relaxation activity. Koi swim in the water features of gardens, restaurants, temples (e.g. Nan Tien Buddhist Temple in Wollongong), offices and public spaces. The species can be seen at institutions like the San Diego Zoo, Smithsonian Institute's National Zoo and Conservation Biology Institute, Sydney's Chinese Gardens and Cowra's Memorial Japanese Gardens.

BREEDING AND GENETICS

Koi colour and scale patterns are influenced by four primary factors: diet, hormones, environment and genetics. Diet can provide pigments like carotenoids that enhance red or orange hues in koi. Hormones like melatonin and cortisol affect the intensity and patterns of koi colouration by regulating pigment cells known as chromatophores (Luo *et al.* 2021). Environmental factors such as light, temperature and water quality also contribute to the overall appearance of koi. However, genetics remains the most important determinant of a koi's colour formation and inherent patterns (Roberts 2023).

Selective breeding has been fundamental to the koi industry since Japanese farmers began the practice in the early 19th century. This method allows breeders to enhance specific genetic traits over generations by selecting parent fish with desired colours, patterns and body shapes. Through intentional hybridisation and line breeding, breeders look to produce offspring with improved aesthetic qualities (De Kock and Gomelsky 2015).

The genetic makeup, or genotype, of koi is the genetic information passed down to its offspring. A koi receives a gene from each of its parents for every characteristic. This includes genes that influence its pigment cells, or chromatophores. Koi have six types of chromatophores – melanocytes (black/dark brown), xanthophores (yellow), erythrophores (red/orange), iridophores (reflective), leucophores (white) and cyanophores (blue). The combination of these chromatophores creates each koi's unique appearance (De Kock and Gomelsky 2015; Luo *et al.* 2021). The expression of genetic traits (the phenotype) is the observable and desirable colour patterns on the fish.

Even with a sound understanding of koi genetics and careful cross-breeding of adult fish, predicting the expression and appearance of inherited colour traits remains challenging. This is because many genes interact in a complex way, and do not always behave as expected (Perelberg *et al.* 2005). Additionally, some genes are recessive (hidden if paired with a dominant gene) or influenced by external conditions, making consistent results difficult to achieve (Tamadachi 1994). As a result, offspring display a wide variety of colours and quality.

Following the breeding process, a rigorous culling procedure selects koi offspring with desirable traits. A high percentage of koi fry is culled regularly at various stages before fish reach 6 months old, often due to unappealing colours or genetic defects (De Kock and

Gomelsky 2015). Show winners are unique individuals selected from hundreds of thousands of candidates.

If you are seeking to delve deeper into koi genetics, advanced texts such as Boris Gomelsky's *Koi Genetics* (Gomelsky 2016) and Lior David's *Inheritance of Colors and Evolution of Koi and Carp* (David 2009) explore the complexities of koi breeding, including the inheritance of colour and other traits. Both authors are respected fish geneticists with decades of experience and they have published extensively in this field. The Koi Organisation International (Pattist 2016) provides an overview of fish genetics, focusing on desirable trait inheritance and advances in genetic understanding and applications in koi aquaculture. For a more interactive experience, computer games like *Pond: A Koi Breeding Game* (https://koifarmgame.com/) allow players to explore koi patterns through selective breeding and engage with an online community of enthusiasts.

Judging koi

Koi value is determined by qualities that breeders aim to enhance through parentage and selection criteria (Tamadachi 1994; De Kock and Gomelsky 2015). Colour and pattern are key factors in the marketability of mass-produced commercial koi, with breeders creating lines that consistently exhibit these traits. In contrast, show koi are evaluated through a point scoring system, always judged when viewed from above. Body shape and size together account for half the score and hold the most importance, with the ideal shape being torpedo-like with well-formed fins. Colour quality, durability, distribution and pattern symmetry are the next most important features. Swimming style is also judged on 'quality', while 'elegance' and 'imposing appearance' each also contribute to the score.

KOI IN AUSTRALIA

Koi organisations, such as the Australian Koi Association and Koi Society of Australia, both based in New South Wales, and the Koi Society of Western Australia, exist for enthusiasts to share knowledge and support one another in koi keeping. The Australian Koi Association (AKA, https://www.akakoi.com.au/) was established in 1981 and has around 300 members. The organisations host regular meetings, shows and auctions, often featuring international koi judges and expert lecturers. For instance, the Koi Society of Australia's Annual Koi Show (http://www.ksakoi.com), held at Fairfield Showground, attracts thousands of visitors each year and features exhibitors from various fields. The Fairfield City Council regards the event as a major tourist attraction for the district. The clubs are family-oriented and multicultural. They provide advisory services, organise events and invest in equipment, like show bins, trailers and pumps. Additionally, they have newsletters, libraries, videos and club magazines to support koi hobbyists and to guide beginners (Fig. 12.3). They work together and have relationships with other Australian and international koi organisations.

Notably, due to the classification of koi as feral species in Australia, koi organisations (e.g. https://www.koiclubwa.com/) and retailers also give advice on environmental issues, disease control, government regulations and humane disposal of koi. For example, the Fish Works Koi Farm (Sydney) provides a New South Wales government brochure urging people to humanely

Fig. 12.3. Examples of newsletters provided by Australia koi organisations. Left to right: Australian Koi Association, Koi Society of Australia and Koi Society of Western Australia.

The Koi Society of Australia Inc.

AUSTRALIAN CHAPTER OF ZEN NIPPON AIRINKAI (ZNA)
Japan
Patron: His Excellency Consul General of Japan
ABN: 11009153457

Member Code of Ethics

The Koi Society of Australia (the Society) is dedicated to "the protection, preservation and improvement of Nishikigoi (Koi) within Australia".

In accordance with this mission statement, any person becoming a member of the Society is therefore expected to:

- Familiarise themselves with the environmental needs of Koi as to housing, feeding, general well being and healthcare in order to provide a safe and humane standard of housing for these living creatures.

- Ensure that safe and humane housing for Koi is maintained by willingly assisting members, particularly novice members, with the sharing of their experiences in keeping Koi.

- Dispose of sick or injured Koi in a humane manner by reference to The procedures outlined in the Society's Handbook.

- Foster and promote the hobby of Koi keeping in Australia in accordance with the constitution of the Society.

- Not do anything that would bring *the Society* or the hobby of keeping Koi into disrepute.

- Operate their Koi infrastructure (ponds, pumps and filters) in an acceptable manner so as to avoid any nuisance to neighbours or impact on the environment.

- Be aware of and abide by the water conservation objectives of the New South Wales Government, including the installation of water saving devices recommended by Sydney Water.

- Familiarise themselves with and abide by the "Rules and Regulations" of the New South Wales Fisheries Department regarding the keeping of Koi and notably as to size of their ponds and disposal of surplus Koi.

- Contemplate and undertake breeding only in order to strive to improve the quality of Australian Koi.

Fig. 12.4. Koi Society of Australia's Code of Ethics. Source: Koi Society of Australia.

rehome unwanted pet fish. It states: 'Your good friend doesn't deserve to be dumped. Rehome pet fish with friends, not in the local river' (NSW Department of Primary Industries nd).

Koi organisations have established codes of ethics and educate their members about responsible practices for the trading and disposal of koi in line with government regulations (Fig. 12.4). Regulated koi auctions, approved by state authorities, create a supervised marketplace for trading koi, discouraging black market activities and preventing environmentally harmful disposal methods (Australian Koi Association Inc. 2006).

A visit to the AKA Koi Auction, October 2024

At the AKA Koi Auction at Fairfield Markets, visitors walk through market stalls before arriving at the koi auction area, which is often held alongside other events. A sign reading 'Enjoy life, Enjoy koi' welcomes attendees (Fig. 12.5). Prior to the auction there is a 30-minute viewing period, where the public, koi breeders and enthusiasts can observe the fish, interact with each other and ask questions. Serious buyers come prepared with spreadsheets and notebooks, planning to buy multiple koi.

Rows of numbered black tubs, lined in blue to highlight the koi's vibrant colours and patterns, hold fish of different sizes. Participants take notes on their preferred koi, and groups of smaller fish typically sell for less than individual large koi. Mixed batches of smaller fish often sell for around $150 for six or eight fish, since smaller koi may change colour with age.

During the auction, the auctioneer, drawing on experience, identifies each koi type. They suggest a starting bid, emphasising stand-out features like solid patterns, desired varieties such as *kohaku*, and ideal body shapes. A camera projects images from fish tub to audience and bidding begins, with top koi fetching over $1000. Once sold, koi are poured into oxygenated bags for collection.

The auction is also an opportunity for newcomers to join the koi club, meet fellow enthusiasts and learn about koi care. Members can rehome unwanted fish and learn how to responsibly keep koi in line with fisheries regulations, ensuring they are not released into the wild. Equipment, such as pumps, and fish food are also available for purchase. Advice is freely given.

The cost of koi keeping depends on whether you want a prize-winning show fish or a backyard pet. Entry into koi keeping will set you back around $2,000 for pond setups and initial fish (prices start at $6 for a 3 cm fish from local pond supplies or garden centres, but average from $100 to $1,500). Quality individual fish at Australian competitions can sell for up to $2,000, with some koi around $5,000 (Koi Society Australia 2024).

Outstanding koi sell for much higher prices. As no two specimens are alike, the grand champions commonly sell for the equivalent of tens of thousands of Australian dollars. In the UK, more famous breeders advertise fish for sale at prices equivalent to around A$10,000–$50,000 for champion-level quality, and many are listed as 'price on application' (Kodama Koi Farm 2024). The most expensive koi purchase to date is S Legend, a female 39-inch (almost 1 m long), 9-year-old *kohaku* variety that was sold from Kentaro Sakai's fish farm for

Fig. 12.5. Sequence of events at the AKA Koi Auction, October 2024. Top to bottom: A welcome sign to the event; tubs for koi viewing prior to the auction; a *kohaku* koi for sale; and sold koi bagged and ready for collection. Source: Katie Doyle.

US$1.8 million (203 million yen) in October 2018 to Ms Yingying Chung, a koi enthusiast and collector (Kodama Koi Farm 2024).). Sadly, this fish died a year after sale in 2019.

Koi ponds have become a central feature of garden landscaping in some high-end homes, with costs sometimes exceeding $100,000 for intricate designs. Notably, in 2023 NBA legend Shaquille O'Neal spent US$500,000 on a 380,000 L koi pond, which included a large koi named 'Sha-koi O'Neal', later renamed 'Charles Barkley' after his fellow player (Aquascape 2024). In Sydney, an A$7.5 million home listed in 2023 on real estate website domain.com. au featured an indoor koi pond with a boardwalk and waterfall. A Japanese-inspired property in New South Wales, listed for A$8.95 million in 2024, included a zen garden and internal jetty overlooking a 25,000 L koi pond. Globally, Christies International Real Estate showcases luxury homes titled 'Living Jewels: 6 Homes with Koi Ponds' that offer tranquil spaces for relaxation, from Rhode Island to the Bahamas (Carsen 2020).

The koi paradox in Australia

In Australia, koi are generally classified as feral carp – i.e. as a noxious or as controlled species across all states and territories. It is not considered a separate species or variant of the common carp in our rivers. Koi is considered a noxious species under the federal *Environment Protection and Biodiversity Conservation Act 1999* and importation is heavily restricted (Department of Agriculture 2024). New South Wales and Western Australia recognise koi as an ornamental fish, but legislation in other states prohibits possession, sale, transportation and release into the wild, with ownership exceptions for fully contained aquaria or private ponds under strict biosecurity and containment measures. These laws are targeted at preventing ecological damage and protecting native fish species.

*****

CyHV-3 is potentially problematic to the koi industry, with high mortality likely if it is released. Limited control options are available, such as vaccines or resistance. Australia likely is free from the virus (although this is not completely certain), but the koi industry has been concerned about it since the early 2000s (Australian Koi Association, Koi Society of Australia and Koi Society of Western Australia 2006). The 2019 report *Biosecurity Strategy for the Koi (*Cyprinus carpio*) Industry* by Lin and Landos, part of the National Carp Control Plan (NCCP), provides a detailed strategy to mitigate CyHV-3 risks (Lin and Landos 2019). It outlines biosecurity measures across four levels: preventing pathogen entry into koi facilities, containing intra-facility spread, limiting pathogen release, and enhancing biosecurity at auctions and shows. Key recommendations include further research, biosecurity education and offshore pathogen management, with quarantine for imported fish. While the proposed measures could reduce risks, the report concluded that maintaining Australia's CyHV-3-free status offered the highest protection to the koi industry by preventing release of the virus and maintaining Australia's current exotic disease status with respect to this pathogen.

13

The future of carp in Australia

We choose some species as desirable, whether for cultural or culinary reasons.
Other species choose us as desirable for their own biological ends (Noakes 1995).

This book has taken us on a long journey from the origins of the common carp, *Cyprinus carpio*, in Europe and Asia around 3 million years ago to the present day. We have travelled back in time 50,000 years to when carp was a food exploited by humans, to carp's aquaculture in China some 8,000 years ago and to its popularity among Roman legionnaires fighting far from home in a region that was perfect for the prolific production of a fishy food source. We have seen the carp's attractions for Christian monks wanting something to contemplate, catch and consume on meat-free days, and its rise to dominance – in what we have called the Carpocene – as *the* aquaculture fish in central and western Europe and many parts of Asia. We have followed the species' introduction to England, where it became the fish *du jour* for aquaculture (not to mention the 'Queen' or wily 'fox' of sport fish), first in monastery ponds and later – thanks in part to Henry VIII and the Dissolution of the monasteries – for a much broader fish-eating public, as its commercial production and culinary appeal grew. Carp even had supporting roles in Shakespeare's *Hamlet* and *All's Well that Ends Well*, signifying its infiltration by the early 1600s not just into English waterways, but into its culture.

It is no surprise then that, together with its distant relatives the salmons and trouts, and its cyprinid cousins the goldfish, tench and roach, carp was brought to Australia. Indeed, it was one of the earliest and more common fishy imports. But the reasons for its introduction were more complex – with theological, anti-Darwinian and even proto-conservation overtones – than simply a wish to make Australia resemble the 'old country'. Many introductions did not establish populations in the new environment. That is not unusual: Williamson and Fritter (1996) proposed the 'tens rule', which suggested that only 1 in 10 introductions turned up in the wild, that only 1 in 10 of those introductions established populations and only 1 in 10 of those actually became an invasive pest. With so many introductions of carp, the numbers of individuals introduced, the sharing of fish by people wanting to rear them in dams or lakes for food or recreation, the enthusiasm for carp by the various acclimatisation societies, *together with* the species' tolerance to extremes of water conditions, it has proven a highly successful invader in Australia (Koehn 2004). The success of carp has been mirrored in many, many other regions of the world (Weber and Brown 2009) and its rise in aquaculture production continues to this day (FAO 2024b).

Probably one of the more intriguing questions is why the species, which was introduced almost 170 years ago, did not proliferate until the Boolarra Event Horizon of the early 1960s. Other populations – made up of the Prospect and Yanco strains – had, it seems, snoozed away relatively quietly around Sydney and in the rivers and lakes of the Murray–Darling Basin (although they were sufficiently abundant to be commercially fished in places) for decades, before interbreeding with the new interloper likely sparked a tsunami, the ripples of which we are still feeling some 60 years on. We may never have a definitive answer. But the spread of Boolarra carp was more than likely helped by an unhappy confluence of natural and human-induced forces: the rapid transition of rivers into irrigation conduits, the construction of impoundments, increasingly degraded native fish populations, the introduction of new genetic material to existing carp populations *and* some of the largest and prolonged periods of flooding for decades.

It is probably pointless to speculate whether we would be knee-deep in carp if Colin Eric Pratt and Geoff Scott had been less enthusiastic about wanting to farm carp at Boolarra, Victoria in the early 1960s. In our view, if it hadn't been them then, someone would have done something similar a few years down the track. After all, there were multiple enquiries in the 1950s about the potential to farm carp in Victoria (Department of Fisheries and Wildlife 1952–1961) and there have been repeated introductions of carp (and other species) to Tasmania and elsewhere since (Yick *et al.* 2021). It is not necessarily what we want to hear, but the carp that made an appearance in *Hamlet* was in the wings of the Australian theatre, waiting for its cue to come on, regardless of who the stage manager was.

We finish this book by asking what lessons we have learnt over the last 170 years or so since carp was introduced to Australia. First, what have we learnt from the history of carp, its spread around the world and into Australia? Second, what have we learnt about the likely future of carp populations in Australia and their impacts? Third, what have we learnt from attempts to eradicate carp overseas? Finally, and in conclusion, the question on everyone's lips: is there any chance we can get rid of it or is carp in Australia here to stay?

WHAT HAVE WE LEARNT FROM CARP'S HISTORY AND SPREAD?

The common carp is tolerant of a range of aquatic conditions, such that it can thrive where Australian native fishes might fear to swim. We have seen people recognise and exploit this tolerance early in the human-history of carp in Asia and Europe, and more recently as carp was introduced to, and established populations in, all continents except Antarctica. Short of depriving carp of water completely, if the species has an environmental Achille's fin, the weakness is far from obvious.

Although carp can at times breed noisily and vigorously and recruit in mind-boggling numbers, this occurs best in newly flooded vegetation, such as in the floodplains of lakes and rivers. The successful Tasmanian eradications were partly due to understanding the need to keep carp away from potential spawning areas (Diggle *et al.* 2012). But it does appear that drought and drying decimate populations, especially if they last for some years, and it can take time for populations to recover (Forsyth *et al.* 2013). No doubt the loss of habitat for adults is part of the reason, but low flows will limit carp to permanent water and minimise successful spawning and/or recruitment (Crook and Gillanders 2006; Stuart and Jones 2006a; Macdonald *et al.* 2010). We can expect the size of carp populations to continue to wax and wane with wet and dry climate cycles in the future.

Slow-flowing or still water environments are particularly conducive to carp proliferation, especially if permanent. Periodically increased water levels of these water bodies is ideal for carp because it inundates normally terrestrial vegetation, which is great for spawning. There is no shortage of such environments in Australia, including the Murray–Darling Basin. Some are natural (e.g. billabongs and floodplain wetlands), some are artificial (e.g. impoundments and farm dams). Like the changed landscape in western Europe brought about by the proliferation of water mills from medieval times (Hoffmann 1995), or the development of paddy-field farming in China thousands of years earlier (Nakajima *et al.* 2019), the environment in Australia, already well-suited for carp, has become even better with the proliferation of permanent still water bodies since colonisation and especially since irrigated farming took off (Gehrke 1997). Through our actions over the last 170 years or so, in Australia we have made a promising environment for carp even better.

WHAT HAVE WE LEARNT ABOUT THE LIKELY IMPACTS OF CARP IN AUSTRALIA?

Lorenzo Vilizzi and colleagues did a nice job in teasing out country-specific attitudes to carp (and non-native and invasive species listings and legislations) on the one hand, and studies of the impacts of the species on the other (Vilizzi *et al.* 2015). Concern, and research into impacts, have been greatest in Australasia and North America where the species was introduced relatively recently and is highly invasive. This means that impacts to ecosystems are likely to be high and perceptions are correspondingly largely negative. Where the species occurs in its natural, ancestral distribution or has become naturalised after being introduced centuries ago, carp tends to be only minorly invasive and attitudes are largely indifferent or positive. Outside its natural range, where the species has been introduced relatively recently but it is seen as useful for aquaculture food production or angling, even though carp may be invasive and potentially problematic to the regions' aquatic ecosystems, attitudes to the species are mostly indifferent.

Although common carp has at various times been blamed for most of the ills of at least the Murray–Darling Basin's rivers, the fish is only one of many insults that we have hurled at the environment. What is more, while undoubtedly carp, even in relatively low densities, has a detrimental effect on turbidity and macrophytes, there are many unanswered questions about the broader effects of carp on still and closed, let alone flowing and open, ecosystems. Our certainty about the impacts of carp diminishes as we climb the food web, although with the emergence and analysis of long-term datasets this is improving all the time. Nevertheless, we are confident about how carp impacts turbidity, algae and macrophytes, less confident about its impacts on zooplankton and macroinvertebrates, and still have a lot to learn when it comes to fish and water birds. Despite the inevitable and understandable limitations of many Australian studies of the ecological impacts of carp, to our knowledge there is little evidence that carp is a positive addition to our waterways. Having said that, the fact that cormorants, pelicans, Murray cod, turtles and golden perch may eat carp at times, especially in locations where rivers are degraded and native fish in short supply (Doyle 2012), means that they can play functional roles in aquatic food webs.

There are relatively few studies of the economic and societal impacts of carp in Australia (Koehn *et al.* 2000; McLeod 2004; National Carp Control Plan 2022c). The National Carp Control Plan investigated the potential socio-economic impacts of using the cyprinid herpesvirus 3 (CyHV-3), whereas the impacts of the presence of carp on people are much harder to find and often incidental and anecdotal (Frawley *et al.* 2012). First Nations peoples, irrigators, natural resource managers, conservation groups, angler groups and scientists mostly view carp negatively and rank it high on the list of threats to Australian rivers (Koehn *et al.* 2000). When John Koehn, Andrea Brumley and Peter Gehrke's report *Managing the Impacts of Carp* was published in 2000, perceptions of the impacts of carp in our waterways, as related by various groups consulted, were sometimes exaggerated. Educational material was produced around that time to inform the public about what was probably true and what was myth, but in many cases the attractiveness of being able to blame carp for the degradation that people were witnessing all around them was and still is tempting.

WHAT HAVE WE LEARNT FROM ATTEMPTS TO ERADICATE CARP OVERSEAS?

To date, there are only a handful of instances where common carp has been eradicated from water bodies, and those were mostly successful only in limited areas in closed systems such as farm dams, reservoirs or small lakes. This was amply shown early in the Boolarra carp saga in Gippsland, when more than 1,300 farm dams and other small water bodies were poisoned during the Victorian carp-kill program (Hume *et al.* 1983). Carp was also eradicated from many farm dams in Tasmania in the 1970s and in 1980 (Inland Fisheries Commission 1975, 1980) and in South Australia in the 1980s (Hall 1988). And of course, we documented the successful eradication in Tasmania in Chapter 9, where carp was confined to and later eradicated from two highland lakes (Yick *et al.* 2021). There have also been successful eradications overseas, but again at relatively small spatial scales. For example, carp was eradicated from Medina Lake (120 ha) in Spain in 2007 using rotenone (Maceda-Veiga *et al.* 2017) and from Pickerel Lake (281 ha) in Minnesota using rotenone along with the installation of an electric fish barrier in 2009 (Huser *et al.* 2022). The latter remains free of carp, whereas the former experienced a new incursion after flooding. The species was also successfully removed from the 0.86 ha Kranskloof Reservoir in South Africa in 2017, using rotenone (Dalu *et al.* 2020). So, carp can be removed – maybe for large land masses it is optimistic to say 'eradicated' – from relatively small, still water bodies, and with the construction of barriers it can be prevented from re-invading.

As far as we know, however, common carp has never been eradicated from an open system, such as a river, anywhere in the world. In each case that carp was removed from a closed system, it was done by using a poison alone or combined with physical removal, radio-tracking and prevention of access to spawning sites, as in the Tasmanian lakes (Yick *et al.* 2021). If large-scale eradication is desirable, especially from open systems, then something more radical, more far-reaching, is needed.

The deployment of CyHV-3 should in theory be effective in both open and closed systems, as long as fish mix and so can infect each other. It has the potential to be a game-changer, because success is largely reliant on the behaviour of the virus and the fish themselves. The human factor may involve reducing numbers of carp before the virus is released (to reduce the number of dead carp, and the associated water quality issues) and cleaning up afterwards. Neither of these is trivial, but also neither will affect the success of the virus *per se*. Gaining and keeping the support of the general public is another matter. If the virus reduces populations to 40–60% of pre-release levels, as predicted from modelling (National Carp Control Plan 2022a), and environmental conditions create immunity in even a small proportion of fish, then carp might get back to its current numbers within only a few years (Becker *et al.* 2019). Would the initial reduction in carp numbers be enough that other carp-reduction approaches, such as genetic intervention (National Carp Control Plan 2022b), could further reduce numbers such that the environment would significantly benefit?

IS CARP HERE TO STAY?

If we were betting people, we would say that – unless a novel and extremely effective way is found to eradicate carp – the odds are that carp will be in Australia for the foreseeable future. However, CRISPR gene drive technology seems to have promise for dealing with invasive species (e.g. Champer *et al.* 2021). Watch this space! But eradication in open systems has not been achieved to date, let alone eradication in whole catchments, let alone anything the size of the Murray–Darling Basin. Should we expect a different outcome in Australia? Much of the rest of the world, for various reasons, has come to accept and even deliberately embrace carp. For many, carp meets the need for easily produced, high-quality food.

We finish with two questions that in our view have clear answers, and two which we will leave with you. Should we give up trying to eradicate carp? No. In our view, the benefits of completely getting rid of carp from Australia would benefit our freshwater ecosystems enormously, and it is something we should continue to strive for. But we will not hold our collective breaths. Should we in the meantime give up trying to control and manage carp, and so limit its impact? Definitely not. But trading off the perceived and actual impact of carp relative to other environmental stressors is crucial. There are many, many other environmental stressors that we need to fix. And they all need funding. Carp might be able to be removed from a particular water body with enough money, time and effort, but keeping it out on an ongoing basis, especially in the event of periodic flooding, is another thing altogether. Finally, should CyHV-3 be released as a way of knocking numbers of carp down and then the lower numbers be hit hard with complementary measures of removal? Or would the funds needed, which are substantial, be better spent in other ways to control carp? Discuss.

Appendix 1

Carp introductions to Australia before Boolarra.

Source	Year introduced	State	Location	Name of species	Number	Primary or secondary source
The Argus 2 March 1858, p. 4	1858	Tas	Cascade Reservoir, Hobart	Carp		P
Acclimatisation Society of Victoria Annual Report 1 1862, p. 42	1862	Vic	?	English river carp	3	P
Acclimatisation Society of Victoria Annual Report 2 1863, p. 39	1862–1863?	Vic	Royal Park and Botanic Gardens, Melbourne and Phillip Island, Westernport	Carp	7	P
Acclimatisation Society of Victoria Annual Report 3 1864, p. 32	1862–1864?	Vic	Various localities favourable to their multiplication	Carp		P
Acclimatisation Society of Victoria Annual Report 4 1866, p. 9	1865?	Tas		English carp		P
Acclimatisation Society of NSW 4th Annual Report 1865, p. 62	1865–1866	NSW	Lagoons and creeks between Sydney and Botany	Prussian carp		P
Tasmanian Morning Herald 7 May 1866, p. 3	?			Prussian carp		P
Zoological and Acclimatisation Society of Victoria 1872, p. 11		Vic	Wimmera River	Carp		P
Zoological and Acclimatisation Society of Victoria 1873, p. 10	1872	Vic	Reservoirs at Maldon, Sandhurst, Castlemaine, Maryborough etc.	Carp	Large number	P
Zoological and Acclimatisation Society of Victoria 1873, p. 35	1872	Vic	Liberated in the bush	Carp	1,700	P

continued

Source	Year introduced	State	Location	Name of species	Number	Primary or secondary source
Zoological and Acclimatisation Society of Victoria 1874, p. 11	1873	Vic	Liberated in the bush	Carp	Large number, >2,000	P
Zoological and Acclimatisation Society of Victoria 1874, p. 46	1873	Vic	Society's Gardens in Royal Park	Prussian carp		P
Zoological and Acclimatisation Society of Victoria 1875, p. 12	1875	Vic	Lake Colac	Carp	600–700	P
Zoological and Acclimatisation Society of Victoria 1875, p. 32	1874–1875		Liberated	Carp	800	P
Geelong and Western District Fish Acclimatising Society 1875, p. 3		Vic	Ponds	Carp	Some	P
Blackwood Fish Acclimatisation Society. Annual Report 1875, p. 3	1873 or 1874	Vic	Blackwood Reservoir?, Bacchus Marsh?	European carp		P
The Ballarat Star 1876, p. 2	1875	Vic	Lake Wendouree, Ballarat	Carp		
The Australasian 1877, p. 19	1877	Vic	Geelong	Carp	1,000	P
Zoological and Acclimatisation Society of Victoria 1878, p. 127	1878	Vic	Tributaries of Lake Corangamite	Carp	A few	P
The Argus 1888, p. 9	1888	NSW	Upper Murrumbidgee and Snowy Rivers near Yass	Russian carp	300 young trout, perch and carp	P
Reports of the NSW Fisheries Dept, 1916 (in Reid *et al.* 1997)	1916	NSW	Lake Windamere	Prussian carp		P
McCulloch 1929			Introduced to Australia	*Cyprinus carpio*		S
Stead 1929, p. 100	1907	NSW	Prospect Reservoir, in the inlet pond called Bloxsome Pond	Great carp, *Cyprinus carpio*	9 in Prospect Reservoir, 3 kept in aquaria by Stead	P

Source	Year introduced	State	Location	Name of species	Number	Primary or secondary source
Stead 1929, p. 101	1908	NSW	A race in the trout hatchery	Great carp, *Cyprinus carpio*	6	P
Whitley 1951, p. 234	1908	NSW	See Stead (1929) for details. Same fish. And in Taronga Park Aquarium. Still there in 1951	Prussian carp	Some	S
Whitley 1951, p. 234	Before 1920	NSW	Other enclosed waters	Prussian carp	Some	?
Roughley 1951, p. 154	Before 1876	Tas		Common carp, *Cyprinus carpio*		S
Roughley 1951, p. 155	1876	Vic	Geelong and Western District Fish Acclimatising Society distributed carp widely in Victorian streams	Common carp, *Cyprinus carpio*		S
Wiltshire 1962b, p. 11	1907 and 1908	NSW	Prospect Reservoir, later liberated in Lake Windemere and in Centennial Park, Sydney. After Stead, and letter from Whitley	European carp, *Cyprinus carpio*		S
Lake 1967, p. 16	?	?	Widespread, but only common in irrigation canals and sluggish waters in NSW	European, common, German, Asian or edible carp, *Cyprinus carpio*		S
Wharton 1971	1960	Vic	Boolarra, Gippsland	European carp, *Cyprinus carpio*	Some	P
Scott *et al.* 1974, p. 324	About 1890	NSW	Dams	European carp, *Cyprinus carpio*		S
Lake 1978, pp. 79–80	Many decades before 1978			Common carp, *Cyprinus carpio*		S
Brumley 1996, p. 102	1876	NSW	Murrumbidgee River	Singaporean strain of koi		S
Smith 2005, p. 9	1876	NSW	Griffith	*Cyprinus carpio*		S

References

Acclimatisation Society of Victoria (1862) *The First Annual Report of the Acclimatisation Society of Victoria: With the Addresses Delivered at the Annual Meeting of the Society held November 24th 1862 at the Mechanics' Institute Melbourne, by His Excellency Sir Henry Barkly, KCB and Professor McCoy*. Wilson and Mackinnon, Melbourne.

Acclimatisation Society of Victoria (1864) *Annual Report of the Acclimatisation Society of Victoria: With the Addresses Delivered at the Annual Meeting of the Society … 1st–7th (1862–1871)*. Wilson and Mackinnon, Melbourne.

Acclimatisation Society of Victoria (1866) *Annual Report of the Acclimatisation Society of Victoria: With the Addresses Delivered at the Annual Meeting of the Society … 1st–7th (1862–1871)*. Wilson and Mackinnon, Melbourne.

Adamek Z (1998) Breeding biology of carp (*Cyprinus carpio* L.) in the Murrumbidgee Irrigation Area. Visiting Scientists Report. CSIRO Division of Land and Water, Griffith, NSW.

Adzhimuradov K (1972) The food of juvenile carp (*Cyprinus carpio* (L.)) in early development stanzas in bodies of water of the Arakum (Terek River Delta). *Journal of Ichthyology* **12**, 981–986.

Akhurst DJ, Jones GB, Clark M, Reichelt-Brushett A (2012) Effects of carp, gambusia, and Australian bass on water quality in a subtropical freshwater reservoir. *Lake and Reservoir Management* **28**(3), 212–223. doi:10.1080/07438141.2012.702337

Amano M (1968) *Colourful 'Live Jewels': General Survey of Fancy Carp*. Kajima Shoten Publishing, Tokyo.

American Carp Society (2020) *Carp Care 'Stewardship for the Future'* [online]. Available: <https://americancarpsociety.com/carp-care#:~:text=Wet%20the%20sling%20down%20and,after%20un%2Dhooking%20with%20efficiency> (accessed 16 January 2024).

Andersen TB, Jensen PS, Skovsgaard CV (2016) The heavy plow and the agricultural revolution in medieval Europe. *Journal of Development Economics* **118**, 133–149. doi:10.1016/j.jdeveco.2015.08.006

Anderson H (1918) Rescue operations on the Murrumbidgee River. *Proceedings of the Royal Zoological Society of NSW* **1**, 157–160.

Anderson W (1992) Climates of opinion: acclimatization in nineteenth-century France and England. *Victorian Studies* **35**(2), 135–157.

AnglingTimes (2018) Surface fishing tips, rigs and bait for carp. Available: <https://www.anglingtimes.co.uk/advice/tips/surface-fishing-tips-rigs-and-bait-for-carp/> (accessed 27 July 2024).

Anon (1714) The whole art of fishing. Being a collection and improvement of all that has been written upon this subject: with many new experiments. Shewing the different ways of angling, and the best methods of taking fresh-water fish. To which is added, The laws of angling. E. Curll, London.

Anon (2010) Origins & Myths. *'Waddy' Koi Kichi … since 1972* [online]. Available: <http://koikichi.com/origins-myths/> (accessed 12 December 2024).

Anon (2017) Secrets of Fly Fishing the Lowly Carp. *Colorado Trout Unlimited* [online]. Available: <https://coloradotu.org/blog/2017/05/secrets-of-fly-fishing-the-lowly-carp> (accessed 7 May 2024).

Aquascape (2024) *Shaquille O'Neal: Artists of the Year Collaboration* [online]. Available: <https://www.aquascapeinc.com/artistbook/project/shaq/> (accessed 1 November 2024).

Arlinghaus R, Mehner T (2003) Socio-economic characterisation of specialised common carp (*Cyprinus carpio* L.) anglers in Germany, and implications for inland fisheries management and eutrophication control. *Fisheries Research* **61**(1–3), 19–33. doi:10.1016/S0165-7836(02)00243-6

Arlinghaus R, Aas Ø, Alós J, Arismendi I, Bower S, *et al.* (2021) Global participation in and public attitudes toward recreational fishing: international perspectives and developments. *Reviews in Fisheries Science & Aquaculture* **29**(1), 58–95. doi:10.1080/23308249.2020.1782340

Armstrong T (2025) Eat the invaders. *Episode 2: Carp*. ABC.

Ashcraft B, Benny H (2016) *Japanese Tattoos: History *Culture *Design*. Tuttle Publishing, Tokyo.

Australian Academy of Science (2019) *Investigation of the Causes of Mass Fish Kills in the Menindee Region NSW over the Summer of 2018-2019*. Australian Academy of Science, Canberra.

Australian and New Zealand Environment and Conservation Council (2000) *Australian and New Zealand Guidelines for Fresh and Marine Water Quality*. Australian Water Association, Sydney.

Australian Fisheries Newsletter (1965) European carp spread. *Australian Fisheries Newsletter* **24**(11), 17.

Australian Fisheries Newsletter (1970) Noxious carp caught in River Murray. *Australian Fisheries Newsletter* **29**, 29.

Australian Fisheries Newsletter (1976a) Fish tagging in Murray brings in interesting results. *Australian Fisheries Newsletter* **35**, 29.

Australian Fisheries Newsletter (1976b) Fishermen oppose clubs' carp plan. *Australian Fisheries Newsletter* **35**, 25.

Australian Koi Association, Koi Society of Australia, Koi Society of Western Australia (2006) Joint Submission to the Ornamental Fish Policy Working Group, Department of Agriculture, Fisheries and Forestry. Available: <https://www.ksakoi.com/koisub.pdf> (accessed 29 July 2025).

Axelrod HR (1973) *Koi of the World: Japanese Colored Carp*. TFH Publications, Neptune City, NJ.

Axelrod HR (1988) *Koi Varieties: Japanese Colored Carp-Nishikigoi*. TFH Publications, Neptune City, NJ.

Badiou P, Goldsborough GL, Wrubleski D (2011) Impacts of the common carp (*Cyprinus carpio*) on freshwater ecosystems: a review. In *Carp: Habitat, Management and Diseases*. (Eds JD Sanders, SB Peterson) pp. 121–146. Nova Science Publishers, New York.

Bajer PG, Sorensen PW (2010) Recruitment and abundance of an invasive fish, the common carp, is driven by its propensity to invade and reproduce in basins that experience winter-time hypoxia in interconnected lakes. *Biological Invasions* **12**, 1101–1112. doi:10.1007/s10530-009-9528-y

Bajer PG, Sorensen PW (2015) Effects of common carp on phosphorus concentrations, water clarity, and vegetation density: a whole system experiment in a thermally stratified lake. *Hydrobiologia* **746**, 303–311. doi:10.1007/s10750-014-1937-y

Bajer PG, Chizinski CJ, Silbernagel JJ, Sorensen PW (2012) Variation in native micro-predator abundance explains recruitment of a mobile invasive fish, the common carp, in a naturally unstable environment. *Biological Invasions* **14**(9), 1919–1929. doi:10.1007/s10530-012-0203-3

Bajer PG, Parker JE, Cross TK, Venturelli PA, Sorensen PW (2015) Partial migration to seasonally-unstable habitat facilitates biological invasions in a predator-dominated system. *Oikos* **124**(11), 1520–1526. doi:10.1111/oik.01795

Balon EK (1995a) The common carp, *Cyprinus carpio*: its wild origin, domestication in aquaculture, and selection as colored nishikigoi. *Guelph Ichthyology Reviews* **3**, 1–55.

Balon EK (1995b) Origin and domestication of the wild carp, *Cyprinus carpio*: from Roman gourmets to the swimming flowers. *Aquaculture* **129**(1–4), 3–48. doi:10.1016/0044-8486(94)00227-F

Balon EK (2004) About the oldest domesticates among fishes. *Journal of Fish Biology* **65**(s1), 1–27. doi:10.1111/j.0022-1112.2004.00563.x

Balon EK (2006) The oldest domesticated fishes, and the consequences of an epigenetic dichotomy in fish culture. *Aqua: Journal of Ichthyology & Aquatic Biology* **11**(2), 47–87.

Banet NV, Fieberg J, Sorensen PW (2022) Migration, homing and spatial ecology of common carp in interconnected lakes. *Ecology of Freshwater Fish* **31**(3), 164–176. doi:10.1111/eff.12622

Bank Angler (2024) *Bank Angling* [online]. Available: <https://www.bankangler.com.au/bank-angling.html#/> (accessed 11 December 2023).

Bank Angler (2025) Bank Angler Trans-Tasman Competition. Available: <https://www.bankangler.com.au/about---batt.html#/> (accessed 24 July 2025).

Barrett J (2003) Australia's national management strategy for carp control. Managing invasive freshwater fish in New Zealand. *Proceedings of a Workshop Hosted by Department of Conservation*, 2003. Department of Conservation, pp. 155–170.

Barriault I, Barabe D, Cloutier L, Pellerin S, Gibernau M (2021) Pollination ecology of *Symplocarpus foetidus* (Araceae) in a seasonally flooded bog in Quebec, Canada. *Botany Letters* **168**(3), 373–383. doi:10.1080/23818107.2021.1909496

Bartz B (2017) Ghost koi: what the heck is a ghostie? *Koi Organisation International* [online]. Available: <https://koiorganisationinternational.org/blog-entry/ghost-koi-what-heck-ghostie> (accessed 16 April 2024).

Baxter CV, Fausch KD, Saunders WC (2005) Tangled webs: reciprocal flows of invertebrate prey link streams and riparian zones. *Freshwater Biology* **50**(2), 201–220. doi:10.1111/j.1365-2427.2004.01328.x

Becker JA, Ward MP, Hick PM (2019) An epidemiologic model of koi herpesvirus (KHV) biocontrol for carp in Australia. *Australian Zoologist* **40**(1), 25–35. doi:10.7882/AZ.2018.038

Benecke N (1986) Some remarks on sturgeon fishing in the southern Baltic region in medieval times. Fish and Archaeology. *BAR International Series* **294**, 9–17.

Berg LS (1965) *Freshwater Fishes of the U.S.S.R. and Adjacent Countries.* Israel Program for Scientific Translations, Jerusalem.

Bergmann SM, Jin Y, Franzke K, Grunow B, Wang Q, *et al.* (2020) Koi herpesvirus (KHV) and KHV disease (KHVD): a recently updated overview. *Journal of Applied Microbiology* **129**(1), 98–103. doi:10.1111/jam.14616

Berlage E (2022) Celebrity chefs say carp recipes could help clear environmental menace. ABC.

Berra TM (2001) *Freshwater Fish Distribution.* Academic Press, San Diego.

Beukma J (1970) Angling experiments with carp (*Cyprinus carpio* L.). *Netherlands Journal of Zoology* **19**, 81–92.

Bomford M, Tilzey R (1997) Pest management principles for European carp. In *Controlling Carp: Exploring the Options for Australia. Proceedings of a Workshop*, 22–24 October 1996, Albury, NSW. (Eds J Roberts, R Tilzey) pp. 9–20. CSIRO Land and Water, Griffith, NSW.

Bond J (2016a) Fishponds in the monastic economy in England. In *Historical Aquaculture in Northern Europe.* (Eds M Bonow, H Olsen and I Svanberg) pp. 29–57. Södertörn University, Södertörn, Sweden.

Bond J (2016b) The increase of those creatures that are bred and fed in the water. In *Historical Aquaculture in Northern Europe.* (Eds M Bonow, H Olsen and I Svanberg) pp. 157–199. Södertörn University, Södertörn, Sweden.

Bonow M, Svanberg I (2016) Historical pond-breeding of cyprinids in Sweden and Finland. In *Historical Aquaculture in Northern Europe.* (Eds M Bonow, H Olsen and I Svanberg) pp. 89–119. Södertörn University, Södertörn, Sweden.

Bonow M, Olsen H, Svanberg I (2016a) Fishponds in the Baltic states: historical cyprinid culture in Estonia, Latvia and Lithuania. In *Historical Aquaculture in Northern Europe.* (Eds M Bonow, H Olsen and I Svanberg) pp. 139–156. Södertörn University, Södertörn, Sweden.

Bonow M, Olsén H, Svanberg I (Eds) (2016b) *Historical Aquaculture in Northern Europe.* Södertörn University, Södertörn, Sweden.

Boutier M, Donohoe O, Kopf RK, Humphries P, Becker J, *et al.* (2019) Biocontrol of carp: the Australian plan does not stand up to a rational analysis of safety and efficacy. *Frontiers in Microbiology* **10**, 882. doi:10.3389/fmicb.2019.00882

Bowerman MR (1975) Concern over the spread of carp in southern Australia. *Australian Fisheries Newsletter* **34**, 4–7.

Bowmer K, Bales M, Roberts J (1994) Potential use of irrigation drains as wetlands. *Water Science and Technology* **29**(4), 151–158. doi:10.2166/wst.1994.0179

Breheny K (1996) Introduced cyprinid (carp) fishes in Western Australia and their management implications. Bachelor of Science (Env. Mgt) Hons thesis, Edith Cowan University, Perth.

Britton JR (2023) Contemporary perspectives on the ecological impacts of invasive freshwater fishes. *Journal of Fish Biology* **103**(4), 752–764. doi:10.1111/jfb.15240

Brown AG, Lespez L, Sear DA, Macaire J-J, Houben P, *et al.* (2018) Natural vs anthropogenic streams in Europe: history, ecology and implications for restoration, river-rewilding and riverine ecosystem services. *Earth-Science Reviews* **180**, 185–205. doi:10.1016/j.earscirev.2018.02.001

Brown AM (1980) *An Evaluation of the Role of Genetics in the Management of Victorian Populations of Carp (Cyprinus carpio L.).* Arthur Rylah Institute for Environmental Research, Melbourne.

Brown P (1996) *Carp in Australia.* NSW Fisheries, Sydney.

Brown P, Gilligan D (2014) Optimising an integrated pest-management strategy for a spatially structured population of common carp (*Cyprinus carpio*) using meta-population modelling. *Marine and Freshwater Research* **65**(6), 538–550. doi:10.1071/MF13117

Brown P, Walker TI (2004) CARPSIM: stochastic simulation modelling of wild carp (*Cyprinus carpio* L.) population dynamics, with applications to pest control. *Ecological Modelling* **176**(1–2), 83–97. doi:10.1016/j.ecolmodel.2003.11.009

Brown P, Sivakumaran KP, Stoessel D, Giles A, Green C, *et al.*(2003) *Carp Population Biology in Victoria*, Marine and Freshwater Resources Institute, Department of Primary Industry, Snobs Creek, Victoria.

Brown P, Sivakumaran KP, Stoessel D, Giles A (2005) Population biology of carp (*Cyprinus carpio* L.) in the mid-Murray River and Barmah Forest wetlands, Australia. *Marine and Freshwater Research* **56**(8), 1151–1164. doi:10.1071/MF05023

Brumley A (1991) Cyprinids of Australasia. In *Cyprinid Fishes: Systematics, Biology and Exploitation*. (Eds IJ Winfield, JS Nelson) pp. 264–283. Springer, Dordrecht.

Brumley A (1996) Family Cyprinidae: carps, minnows etc. In *Freshwater Fishes of South-eastern Australia*. 2nd edn. (Ed. RM McDowall) pp. 99–106. Reed Books, Sydney.

Brykała D, Podgórski Z (2020) Evolution of landscapes influenced by watermills, based on examples from northern Poland. *Landscape and Urban Planning* **198**, 103798. doi:10.1016/j.landurbplan.2020.103798

Buckland FT (1861) *The Acclimatisation of Animals: A Paper*. Acclimatisation Society of Victoria, Melbourne.

Buckmaster D (1975) *Situation Report: European Carp*. Fisheries and Wildlife Division, Ministry for Conservation Victoria.

Buffler R, Dickson T (2009) *Fishing for Buffalo: A Guide to the Pursuit and Cuisine of Carp, Suckers, Eelpout, Gar, and Other Rough Fish*. University of Minnesota Press, Minneapolis.

Bullers RM (2024) Acoustic conditioning in common carp (*Cyprinus carpio*) to accelerate removal and reduce cost. MS, University of Minnesota, Minneapolis.

Bunn SE, Arthington AH (2002) Basic principles and ecological consequences of altered flow regimes for aquatic biodiversity. *Environmental Management* **30**(4), 492–507. doi:10.1007/s00267-002-2737-0

BushnBeachFishing (2023) *2023 Wyaralong Dam Carp and Tilapia Fishing Competition* [online]. Available: <https://bnbfishing.com.au/2023-wyaralong-dam-carp-and-tilapia-fishing-competition/> (accessed 12 December 2023).

Butcher AD (1962) *Why Destroy the European Carp?* Fisheries and Wildlife Department, Melbourne.

Bzonek PA (2017) Acoustic and strobe-light behavioural barriers: examining behavioural and movement responses of common carp (*Cyprinus carpio*) at laboratory and mesocosm scales. MSc, University of Toronto, Canada.

Bzonek PA, Kim J, Mandrak NE (2020) Short-term behavioural response of common carp, *Cyprinus carpio*, to acoustic and stroboscopic stimuli. *Management of Biological Invasions* **11**(2), 279–292. doi:10.3391/mbi.2020.11.2.07

Cadwallader PL (1977) J.O. Langtry's 1949–50 Murray River investigations. Fisheries and Wildlife, Paper 13. Fisheries and Wildlife, Melbourne.

Cahn AR (1929) The effect of carp on a small lake: the carp as a dominant. *Ecology* **10**(3), 271–274. doi:10.2307/1929502

Carl DD, Weber MJ, Brown ML (2016) An evaluation of attractants to increase catch rates and deplete age-0 common carp in shallow South Dakota lakes. *North American Journal of Fisheries Management* **36**(3), 506–513. doi:10.1080/02755947.2016.1141125

Carp books online (2024) *Carp Books Online* [online]. Available: <https://www.carpbooksonline.com/> (accessed 23 January 2024).

Carpology (2017) *How to Hold a Carp Correctly* [online]. Available: <https://www.carpology.net/article/other-stuff/how-to-hold-a-carp-correctly/#:~:text=Place%20your%20index%20finger%20one,mat%20until%20it%20calms%20down> (accessed 15 December 2023).

Carpology (2020) The 25 greatest captures that shaped carp fishing: Part 1 and Part 2. *Carpology, Other Stuff* [online]. Available: <https://www.carpology.net/article/other-stuff/the-25-greatest-captures-that-shaped-carp-fishing-part-1/> (accessed 5 October 2023).

Carpworld (2023) Available: <https://carpworld.co.uk/>.

Carsen L (2020) Living jewels: 6 homes with koi ponds. *Luxury Defined by Christie's International Real Estate* [online]. Available: <https://www.christiesrealestate.com/blog/living-jewels-koi-ponds-and-water-gardens/> (accessed 18 June 2024).

Carus-Wilson E (2017) An industrial revolution of the thirteenth century 1. In *Internal Colonization in Medieval Europe*. (Eds F Fernandez-Armesto, J Muldoon) pp. 301–322. Routledge, London.

Casal CMV (2006) Global documentation of fish introductions: the growing crisis and recommendations for action. *Biological Invasions* **8**(1), 3–11. doi:10.1007%2fs10530-005-0231-3

Casas L, Szűcs R, Vij S, Goh CH, Kathiresan P, Nemeth S, Jeney Z, Bercsenyi M, Orban L (2013) Disappearing scales in carps: re-visiting Kirpichnikov's model on the genetics of scale pattern formation. *PLOS One* **8**, e83327.

Cavender T, Coburn M (1992) Phylogenetic relationships of North American Cyprinidae. In *Systematics, Historical Ecology, & North American Freshwater Fishes*. (Ed. RL Mayden) pp. 293–327. Stanford University Press, Redwood City, CA.

Cavender TM (1991) The fossil record of the Cyprinidae. In *Cyprinid Fishes: Systematics, Biology and Exploitation*. (Eds IJ Winfield, JS Nelson) pp. 34–54. Springer, Edmonton.

Centre for Invasive Species Solutions (2012) Overview of daughterless carp research. Factsheet. PestSmart website: <https://pestsmart.org.au/toolkit-resource/overview-of-daughterless-carp-research>.

Centre for Invasive Species Solutions (2014) Fishing as a carp control technique. Factsheet. PestSmart website: <https://pestsmart.org.au/toolkit-resource/fishing-as-a-carp-control-method>.

Champer SE, Oakes N, Sharma R, García-Díaz P, Champer J, *et al.* (2021) Modeling CRISPR gene drives for suppression of invasive rodents using a supervised machine learning framework. *PLoS Computational Biology* **17**(12), e1009660. doi:10.1371/journal.pcbi.1009660

Chen D, Zhang Q, Tang W, Huang Z, Wang G, *et al.* (2020) The evolutionary origin and domestication history of goldfish (*Carassius auratus*). *Proceedings of the National Academy of Sciences* **117**(47), 29775–29785. doi:10.1073/pnas.2005545117

Cherrell K (2022) *Two Tone: Funeral for a Fish* [online]. Available: <https://burialsandbeyond.com/2022/10/01/two-tone-funeral-for-a-fish/> (accessed 12 December 2023).

ChinaOnlineMuseum (2024) *Wu Qingxia: Nine Carps* [online]. Available: <https://www.comuseum.com/product/wu-qingxia-nine-carps/> (accessed 1 November 2024).

Chistiakov DA, Voronova NV (2009) Genetic evolution and diversity of common carp *Cyprinus carpio* L. *Central European Journal of Biology* **4**(3), 304–312. doi:10.2478/s11535-009-0024-2

Cios S (2016) The history of aquaculture in Poland. In *Historical Aquaculture in Northern Europe*. (Eds M Bonow, H Olsen and I Svanberg) pp. 59–76. Södertörn University, Södertörn, Sweden.

Clements J (1988) *Salmon at the Antipodes: a History and Review of Trout, Salmon and Char and Introduced Coarse Fish in Australasia*. Published by the author, Ballarat, Victoria.

Clifford K (1992) *A History of Carp Fishing*. Sandholme Publishing, Newport.

CoastFM (2023) *Koroit Carp Fishing Comp 2023* [online]. Available: <https://coast.com.au/events/koroit-carp-fishing-comp/> (accessed 2 February 2023).

Colledge S, Conolly J (2007) *The Origins and Spread of Domestic Plants in Southwest Asia and Europe*. Left Coast Press, Walnut Creek, CA.

Colvin ME, Pierce CL, Stewart TW, Grummer SE (2012) Strategies to control a common carp population by pulsed commercial harvest. *North American Journal of Fisheries Management* **32**(6), 1251–1264. doi:10.1080/02755947.2012.728175

Conallin AJ, Smith BB, Thwaites LA, Walker KF, Gillanders BM (2012) Environmental water allocations in regulated lowland rivers may encourage offstream movements and spawning by common carp,

Cyprinus carpio: implications for wetland rehabilitation. *Marine and Freshwater Research* **63**(10), 865–877. doi:10.1071/MF12044

Conallin AJ, Smith BB, Thwaites LA, Walker KF, Gillanders BM (2016) Exploiting the innate behaviour of common carp, *Cyprinus carpio*, to limit invasion and spawning in wetlands of the River Murray, Australia. *Fisheries Management and Ecology* **23**(6), 431–449. doi:10.1111/fme.12184

Context (2019) Latrobe Valley social history: celebrating and recognising Latrobe Valley's history and heritage. Victoria Department of Environment, Land, Water and Planning.

Corby J (2022) 20 astounding koi fish tattoo designs with meaning. *Tattoo Like the Pros* [online]. Available: <https://tattoolikethepros.com/koi-fish-tattoo/> (accessed 1 October 2024).

Costar BJ (2002) Rylah, Sir Arthur Gordon (1909–1974). *Australian Dictionary of Biography*. Australian National University, Canberra.

Crivelli A (1981) The biology of the common carp, *Cyprinus carpio* L. in the Camargue, southern France. *Journal of Fish Biology* **18**(3), 271–290. doi:10.1111/j.1095-8649.1981.tb03769.x

Crook DA (2004) Is the home range concept compatible with the movements of two species of lowland river fish? *Journal of Animal Ecology* **73**(2), 353–366. doi:10.1111/j.0021-8790.2004.00802.x

Crook DA, Gillanders BM (2006) Use of otolith chemical signatures to estimate carp recruitment sources in the mid-Murray River, Australia. *River Research and Applications* **22**(8), 871–879. doi:10.1002/rra.941

Crosby AW (2000) *Ecological Imperialism: The Biological Expansion of Europe, 900–1900*. Cambridge University Press, Cambridge.

Cummings V, Morris J (2022) Neolithic explanations revisited: modelling the arrival and spread of domesticated cattle into neolithic Britain. *Environmental Archaeology* **27**(1), 20–30. doi:10.1080/14614103.2018.1536498

Currie CK (1989) *Medieval Fishponds in England: Aspects of Their Origin, Function, Management and Development*. University of London, London.

Currie CK (1990) Fishponds as garden features, c. 1550–1750. *Garden History* 18(1), 22–46. doi:10.2307/1586978

Currie CK (1991) The early history of the carp and its economic significance in England. *Agricultural History Review* **39**, 97–107.

Cyr H, Phillips N, Butterworth J (2017) Depth distribution of the native freshwater mussel (*Echyridella menziesii*) in warm monomictic lakes: towards a general model for mussels in lakes. *Freshwater Biology* **62**(8), 1487–1498. doi:10.1111/fwb.12964

Dalu T, Bellingan TA, Gouws J, Impson ND, Jordaan MS, *et al.* (2020) Ecosystem responses to the eradication of common carp *Cyprinus carpio* using rotenone from a reservoir in South Africa. *Aquatic Conservation: Marine and Freshwater Ecosystems* **30**(12), 2284–2297. doi:10.1002/aqc.3463

Damaschin A (2021) Nishikigoi (koi): the shining jewels of Japan. *Japan Creative Enterprise* [online]. Available: <https://jbr.japancreativeenterprise.jp/2021/01/28/nishikigoi-koi-the-shining-jewels-of-japan/> (accessed 28 January 2024).

David L (2009) *Inheritance of Colors and Evolution of Koi and Carp: A Study about Genetic Diversity of Carp Populations, Inheritance of Color Traits in Koi and Evolution of the Tetraploid Genome of this Species*. VDM Verlag Dr. Müller, The Netherlands.

Davies M (1989) The history of nishikigoi. In *The Tetra Encyclopedia of Koi*. (Ed. A McDowall) pp. 10–13. Tetra Press, Morris Plains, NJ.

Davies P, Lawrence S (2019) *Sludge: Disaster on Victoria's Goldfields*. La Trobe University Press, Melbourne.

Davies P, Lawrence S, Turnbull J, Rutherfurd I, Grove J, *et al.* (2018) Reconstruction of historical riverine sediment production on the goldfields of Victoria, Australia. *Anthropocene* **21**, 1–15. doi:10.1016/j.ancene.2017.11.005

Davies PE, Harris JH, Hillman TJ, Walker KF (2010) The sustainable rivers audit: assessing river ecosystem health in the Murray–Darling Basin, Australia. *Marine and Freshwater Research* **61**(7), 764–777. doi:10.1071/MF09043

Davis CC (1959) Damage to fish fry by cyclopoid copepods. *Ohio Journal of Science* **59**, 101–102.

Davis KM, Dixon PI, Harris JH (1999) Allozyme and mitochondrial DNA analysis of carp, *Cyprinus carpio* L., from south-eastern Australia. *Marine and Freshwater Research* **50**(3), 253–260. doi:10.1071/MF97256

Davis P (1975) Noxious carp caught in River Murray. *Australian Fisheries Newsletter* **34**, 28.

Dayman I (2017) 'We're still shaking our heads': SA Carp Frenzy reels in 16,000 fish in one day. ABC Riverland.

De Kock S, Gomelsky B (2015) Japanese ornamental koi carp: origin, variation and genetics. In *Biology and Ecology of Carp*. (Eds C Pietsch, P Hirsch) pp. 27–53. CRC Press, Boca Raton.

Department of Agriculture (2024) *BICON Australian Biosecurity Import Conditions* [online]. Available: <https://bicon.agriculture.gov.au/BiconWeb4.0/> (accessed 16 December 2024).

Department of Fisheries and Wildlife (1952–1961) Commercial Carp Farming. Public Records Office of Victoria, index no. VPRS 12011/P0001.

Derbyshire D (2011) The fat lady sings for the last time: anglers mourn as Britain's largest freshwater fish is found dead. Daily Mail Australia.

Diamond J (2002) Evolution, consequences and future of plant and animal domestication. *Nature* **418**(6898), 700–707. doi:10.1038/nature01019

Didham RK, Tylianakis JM, Hutchison MA, Ewers RM, Gemmell NJ (2005) Are invasive species the drivers of ecological change? *Trends in Ecology & Evolution* **20**(9), 470–474. doi:10.1016/j.tree.2005.07.006

Diggle J, Patil J, Wisniewski C (2012) *A Manual for Carp Control: The Tasmanian Model*. Invasive Animals Cooperative Research Centre, Canberra.

Donkers P (2003a) *An Investigation into the Abundance of European Carp (Cyprinus carpio) in Lakes Crescent and Sorell by the Inland Fisheries Service Carp Management Program*. Inland Fisheries Service, Hobart.

Donkers P (2003b) An Investigation into the Abundance of European Carp (*Cyprinus carpio*) in Lakes Sorell and Crescent. Technical Report.

Donkers P (2004) Age, Growth and Maturity of European carp (*Cyprinus carpio*) in Lakes Sorell and Crescent. Technical Report 18.

Donkers P, Patil J, Wisniewski C, Diggle J (2012) Validation of mark-recapture population estimates for invasive common carp, *Cyprinus carpio*, in Lake Crescent, Tasmania. *Journal of Applied Ichthyology* **28**, 7–14. doi:10.1111/j.1439-0426.2011.01887.x

Doyle KE (2012) Can native Australian percichthyid fishes control invasive common carp (*Cyprinus carpio*)? PhD thesis, University of Queensland, Brisbane.

Driver PD, Harris JH, Norris RH, Closs GP (1997) The role of the natural environment and human impacts in determining biomass densities of common carp in New South Wales rivers. In *Fish and Rivers in Stress: The NSW Rivers Survey*. (Eds JH Harris, PC Gehrke) pp. 225–250.

Driver PD, Closs GP, Koen T (2005a) The effects of size and density of carp (*Cyprinus carpio* L.) on water quality in an experimental pond. *Archiv für Hydrobiologie* **163**(1), 117–131. doi:10.1127/0003-9136/2005/0163-0117

Driver PD, Harris JH, Closs GP, Koen TB (2005b) Effects of flow regulation on carp (*Cyprinus carpio* L.) recruitment in the Murray–Darling Basin, Australia. *River Research and Applications* **21**(2–3), 327–335. doi:10.1002/rra.850

Dyer C (2003) *Everyday Life in Medieval England*. Bloomsbury, London.

Easton C, Elder J (1997) *A Mouthful of Ideas for Better Rivers*. Department of Land and Water Conservation, Tamworth, NSW.

Eaton JG, Scheller RM (1996) Effects of climate warming on fish thermal habitat in streams of the United States. *Limnology and Oceanography* **41**(5), 1109–1115. doi:10.4319/lo.1996.41.5.1109

Ebner B (2006) Murray cod an apex predator in the Murray River, Australia. *Ecology of Freshwater Fish* **15**(4), 510–520. doi:10.1111/j.1600-0633.2006.00191.x

Edwards EA, Twomey K (1982) *Habitat Suitability Index Models: Common Carp*. Fisheries and Wildlife Service, Washington.

Elkins A, Barrow R, Rochfort S (2009) Carp chemical sensing and the potential of natural environmental attractants for control of carp: a review. *Environmental Chemistry* **6**(5), 357–368. doi:10.1071/EN09032

Elkins AC (2021) Bioassay guided fractionation and structural elucidation of environmental attractants for juvenile European carp (*Cyprinus carpio*). PhD thesis, Australian National University, Canberra.

Elkins P (2024) *6 Overlooked, Dependable Carp Flies*. Fly Tyer [online]. Available from: <https://www.flytyer.com/6-overlooked-dependable-carp-flies/>.

Elton CS (2020) *The Ecology of Invasions by Animals and Plants*. Springer Nature, Cham.

Environment Agency (2018) *A Survey of Freshwater Angling in England*. Environment Agency, Bristol.

Eschmeyer WN (2023) *Catalog of Fishes* [online]. Available: <http://www.calacademy.org/research/ichthyology/catalog/fishcatsearch.html> (accessed 8 October 2024).

Fabian MW (1960) Mortality of fresh water and tropical fish fry by cyclopoid copepods. *Ohio Journal of Science* **60**, 268–270.

Fairbairn AS (2000) *Plants in Neolithic Britain and Beyond*. Neolithic Studies Group, Oxford.

Fajt JR, Grizzle JM (1998) Blood respiratory changes in common carp exposed to a lethal concentration of rotenone. *Transactions of the American Fisheries Society* **127**(3), 512–516. doi:10.1577/1548-8659(1998)127<0512:BRCICC>2.0.CO;2

Fanson BG, Hale R, Thiem JD, Lyon JP, Koehn JD, *et al.* (2024) Assessing impacts of a notorious invader (common carp *Cyprinus carpio*) on Australia's aquatic ecosystems: coupling abundance-impact relationships with a spatial biomass model. *Biological Conservation* **290**, 110420. doi:10.1016/j.biocon.2023.110420

FAO (2024a) *Cyprinus carpio Linnaeus 1758* [online]. Food and Agriculture Organisation. Available: <https://www.fao.org/fishery/en/aqspecies/2957/en> (accessed 8 October 2024).

FAO (2024b) *Introduced Speces Fact Sheet: Cyprinus carpio* [online]. Food and Agriculture Organisation. Available: <https://www.fao.org/fishery/en/collection/introsp> (accessed 8 October 2024).

Farag F, Wally Y, Daghash S, Ibrahim A (2014) Some gross morphological studies on the internal anatomy of the scaled common carp fish (*Cyprinus carpio*) in Egypt. *Journal of Veterinary Anatomy* **7**(1), 15–29. doi:10.21608/jva.2014.44724

Faragher R, Lintermans M (1997) Alien fish species from the New South Wales rivers survey. In *Fish and Rivers in Stress: The NSW Rivers Survey*. (Eds JH Harris, PC Gehrke) pp. 201–223. Cooperative Research Centre for Freshwater Ecology, Canberra.

Fellman S, Thomas DM (1987) *The Japanese Tattoo*. Abbeville Press, New York.

Ferguson K (2016) Concerns bow fishing trial could hurt native species. *ABC News*, 12 January 2016.

FineArtAmerica (2024) *Koi Paintings* [online]. Available: <https://fineartamerica.com/art/paintings/koi> (accessed 1 November 2024).

Fisher N, Cribb J (2005) Monitoring community attitudes to using gene technology methods (daughterless carp) for managing common carp. A preliminary investigation. Pest Animal Control Cooperative Research Centre, Canberra.

Fisherman's Forum (1974) Carp in Victorian rivers endanger ecological structure. *Australian Fisheries Newsletter* **33**, 37.

Flores Martin N, Leighton TG, White PR, Kemp PS (2021) The response of common carp (*Cyprinus carpio*) to insonified bubble curtains. *Journal of the Acoustical Society of America* **150**(5), 3874–3888. doi:10.1121/10.0006972

Fornaro J (2020) *Koi Breeds Commonly Misidentified* [online]. Available: <https://hanoverkoifarms.com/mistaken-identity/> (accessed 19 Sepetember 2024).

Forsyth DM, Koehn JD, MacKenzie DI, Stuart IG (2013) Population dynamics of invading freshwater fish: common carp (*Cyprinus carpio*) in the Murray–Darling Basin, Australia. *Biological Invasions* **15**(2), 341–354. doi:10.1007/s10530-012-0290-1

Frasier D (2023) Carp are not bonefish. *Gink and Gasoline* [online]. Available: <https://www.ginkandgasoline.com/carp/carp-are-not-bonefish/> (accessed 30 August 2024).

Frawley J, Nichols S, Goodall H, Baker E (2012) *Talking Fish: Making Connections with the Rivers of the Murray–Darling Basin*. Murray–Darling Basin Authority, Canberra.

Froese R, Pauly D (2023) *FishBase: A Global Information System on Fishes* [online]. Available: <https://www.fishbase.se/search.php> (accessed 12 December 2023).

Frost W (2013) The environmental impacts of the Victorian gold rushes: miners' accounts during the first five years. *Australian Economic History Review* **53**(1), 72–90. doi:10.1111/j.1467-8446.2012.00360.x

Fulton W (2006) The Australian approach to invasive fish species research. The Invasive Asian Carps in North America: A Forum to Understand the Biology and Manage the Problem. Peoria, Illinois.

Furlan EM, Gleeson D, Wisniewski C, Yick J, Duncan RP (2019) eDNA surveys to detect species at very low densities: a case study of European carp eradication in Tasmania, Australia. *Journal of Applied Ecology* **56**(11), 2505–2517. doi:10.1111/1365-2664.13485

Galloway JA (2017) Fishing in medieval England. In *The Sea in History: The Medieval World/La Mer dans L'Histoire: le Moyen Age*. (Ed. M Balard) pp. 629–642. Boydell Press, Martlesham, UK.

García-Berthou E (2001) Size-and depth-dependent variation in habitat and diet of the common carp (*Cyprinus carpio*). *Aquatic Sciences* **63**(4), 466–476. doi:10.1007/s00027-001-8045-6

Geddes M (1979) Salinity tolerance and osmotic behaviour of European carp (*Cyprinus carpio* L.) from the River Murray, Australia. *Transactions of the Royal Society of South Australia* **103**(7), 185–189.

Gehrke P (1997) Differences in composition and structure of fish communities associated with flow regulation in New South Wales rivers. In *Fish and Rivers in Stress: The NSW Rivers Survey*. (Eds JH Harris, PC Gehrke) pp. 169–200. Cooperative Research Centre for Freshwater Ecology, Canberra.

Gehrke P (2001) Preliminary assessment of oral rotenone baits for carp control in New South Wales. In *Managing Invasive Freshwater Fish in New Zealand*. (Ed. R Munro) pp. 143–154. Department of Conservation, New Zealand.

Gehrke PC, Harris JH (1996) Fish and fisheries of the Hawkesbury–Nepean River system. Final Report to the Sydney Water Corporation.

Gehrke PC, Harris JH (2000) Large-scale patterns in species richness and composition of temperate riverine fish communities, south-eastern Australia. *Marine and Freshwater Research* **51**(2), 165–182. doi:10.1071/MF99061

Gehrke PC, Astles KL, Harris JH (1999) Within-catchment effects of flow alteration on fish assemblages in the Hawkesbury–Nepean River system, Australia. *Regulated Rivers: Research & Management* **15**(1–3), 181–198. doi:10.1002/(SICI)1099-1646(199901/06)15:1/3<181::AID-RRR528>3.0.CO;2-5

Gehrke PC, St Pierre S, Matveev V, Clarke M (2010) Ecosystem responses to carp population reduction in the Murray–Darling Basin. *Project MD923 Final Report to the Murray–Darling Basin Authority*. SMEC Australia, Brisbane.

Ghosal R, Coulter AA, Sorensen PW (2022) Proof-of-concept studies demonstrate that food and pheromone stimuli can be used to attract invasive carp so their presence can be readily measured using environmental DNA. *Fishes* **7**(4), 176. doi:10.3390/fishes7040176

Gidmark NJ, Staab KL, Brainerd EL, Hernandez LP (2012) Flexibility in starting posture drives flexibility in kinematic behavior of the kinethmoid-mediated premaxillary protrusion mechanism in a cyprinid fish, *Cyprinus carpio. Journal of Experimental Biology* **215**(Pt 13), 2262–2272. doi:10.1242/jeb.070516

Gillbank L (1986) The origins of the acclimatisation society of Victoria: practical science in the wake of the gold rush. *Historical Records of Australian Science* **6**(3), 359–374. doi:10.1071/HR9860630359

Gillbank L (2001) Animal acclimatisation: McCoy and the menagerie that became Melbourne's zoo. *Victorian Naturalist* **118**(6), 297–304.

Gilligan D, Rayner T (2007) The distribution, spread, ecological impacts and potential control of carp in the upper Murray River. *Fisheries Research Report Series: 14*. New South Wales Department of Primary Industries, Sydney.

Gilligan D, Gehrke PC, Schiller C (2005) Testing methods and ecological consequences of large-scale removal of common carp. NSW Department of Primary Industries & NSW Department of Primary Industries. Narrandera Fisheries Centre.

Gilligan D, Jess L, McLean G, Asmus M, Wooden I., *et al.* (2010) Identifying and implementing targeted carp control options for the Lower Lachlan Catchment. *Industry & Investment NSW – Fisheries Final Report Series No. 118*. NSW Department of Industry & Investment, Batemans Bay, NSW.

Golovanov V, Smirnov A (2007) Influence of the water heating rate upon thermal tolerance in common carp (*Cyprinus carpio* L.) during different seasons. *Journal of Ichthyology* **47**, 538–543. doi:10.1134/s0032945207070089

Gomelsky B (2016) *Koi Genetics*. Amazon Digital Services.

Gozlan RE, Britton J, Cowx I, Copp G (2010) Current knowledge on non-native freshwater fish introductions. *Journal of Fish Biology* **76**(4), 751–786. doi:10.1111/j.1095-8649.2010.02566.x

Graham K, Lowry M, Walford T (2005) *Carp in NSW: Assessment of Distribution, Fishery and Fishing Methods*. NSW Department of Primary Industries, Cronulla Fisheries Centre, Sydney.

Graham K, Gilligan D, Brown P, van Klinken RD, McColl KA, *et al.* (2021) Use of spatio-temporal habitat suitability modelling to prioritise areas for common carp biocontrol in Australia using the virus CyHV-3. *Journal of Environmental Management* **295**, 113061. doi:10.1016/j.jenvman.2021.113061

Grand View Research (2024) *Carp Market Size, Share & Trends Analysis Report by Species (Grass, Silver, Common, Bighead), by Form (Frozen, Canned), by Distribution Channel (Hypermarkets & Supermarkets, Convenience Store), by Region, and Segment Forecasts, 2025–2030* [online]. Available: <https://www.grandviewresearch.com/industry-analysis/carp-market-report> (accessed 16 January 2025).

Hall D (1981) The feeding ecology of the European carp in Lake Alexandrina and the lower River Murray, South Australia. BSc (Hons) thesis, University of Adelaide, Adelaide.

Hall D (1988) The eradication of European carp and goldfish from the Leigh Creek Retention Dam. *Safish* **12**, 15–16.

Hardaker T, Abell J, Chudleigh P, Bennett J, Gillespie R (2019) *Impact Costs of Carp and Expected Benefits & Costs Associated with Carp Control in the Murray–Darling Basin*. Fisheries Research and Development Corporation, Canberra.

Harland J (2019) The origins of aquaculture. *Nature Ecology & Evolution* **3**, 1378–1379. doi:10.1038/s41559-019-0966-3

Harris J (2013) Fishes from elsewhere. In *Ecology of Australian Freshwater Fishes*. (Eds P Humphries, KF Walker) pp. 259–282. CSIRO Publishing, Melbourne.

Harris JH, Gehrke PC (1997) *Fish and Rivers in Stress: The NSW Rivers Survey*. Cooperative Research Centre for Freshwater Ecology, Canberra.

Harvey JH (1992) Westminster Abbey: the infirmarer's garden. *Garden History* **20**(2), 97–115.

Hayes KR, Leung B, Thresher R, Dambacher JM, Hosack GR (2014) Meeting the challenge of quantitative risk assessment for genetic control techniques: a framework and some methods applied to the common carp (*Cyprinus carpio*) in Australia. *Biological Invasions* **16**, 1273–1288.

Haynes G (2009) Population genetics of common carp (*Cyprinus carpio* L.) in the Murray–Darling Basin. PhD thesis, University of Sydney, Sydney.

Haynes G, Gilligan D, Grewe P, Nicholas F (2009) Population genetics and management units of invasive common carp *Cyprinus carpio* in the Murray–Darling Basin, Australia. *Journal of Fish Biology* **75**(2), 295–320. doi:10.1111/j.1095-8649.2009.02276.x

Haynes G, Gilligan D, Grewe P, Moran C, Nicholas F (2010) Population genetics of invasive common carp *Cyprinus carpio* L. in coastal drainages in eastern Australia. *Journal of Fish Biology* **77**(5), 1150–1157. doi:10.1111/j.1095-8649.2010.02742.x

Haynes G, Gongora J, Gilligan D, Grewe P, Moran C, *et al.* (2012) Cryptic hybridization and introgression between invasive Cyprinid species *Cyprinus carpio* and *Carassius auratus* in Australia: implications for invasive species management. *Animal Conservation* **15**(1), 83–94. doi:10.1111/j.1469-1795.2011.00490.x

Hellawell JM (2012) *Biological Indicators of Freshwater Pollution and Environmental Management*. Elsevier, The Netherlands.

Heydarnejad MS (2012) Survival and growth of common carp (*Cyprinus carpio* L.) exposed to different water pH levels. *Turkish Journal of Veterinary & Animal Sciences* **36**(3), 245–249. doi:10.3906/vet-1008-430

Hickley P (1996) Recreational fishing in England and Wales. In *European Inland Fisheries Advisory Commission Report of the Workshop on Recreational Fisheries Planning and Management in Central and Eastern Europe*, Zilinia, Slovakia, 22–25 August 1995: EIFAC Occassional Paper.

Hill MP, Caley P, Stuart I, Duncan RP, Forsyth DM (2024) Large-scale serial replacement of invasive tench (*Tinca tinca*) by invasive carp (*Cyprinus carpio*) in the presence of redfin perch (*Perca fluviatilis*) in the Murray–Darling River system, Australia. *Biological Invasions* **26**(11), 1–17. doi:10.1007/s10530-024-03409-z

Hillyard KA, Smith BB, Conallin AJ, Gillanders BM (2010) Optimising exclusion screens to control exotic carp in an Australian lowland river. *Marine and Freshwater Research* **61**(4), 418–429. doi:10.1071/MF09017

Hillyard KA (2011) Carp exclusion screens on wetland inlets: their value for control of common carp (*Cyprinus carpio* L.) and effects on offstream movements by other fish species in the River Murray, Australia. PhD thesis, University of Adelaide, Adelaide.

Hinds LA, Pech RP (1997) Immuno-contraceptive control for carp. In *Controlling Carp: Exploring the Options for Australia. Proceedings of a Workshop*, 22–24 October 1996, Albury, NSW. (Eds J Roberts, R Tilzey) pp. 108–117. CSIRO Land and Water, Griffith, NSW.

Hoare ME (1966) Learned societies in Australia: the foundation years in Victoria, 1850–1860. *Historical Records of Australian Science* **1**(2), 7–29. doi:10.1071/HR9670120007

Hoda S, Tsukahara H (1971) Studies on the development and relative growth in the carp, *Cyprinus carpio* (Linne). *Journal of the Faculty of Agriculture, Kyushu University* **16**(4), 387–509. doi:10.5109/22811

Hoffmann RC (1995) Environmental change and the culture of common carp in medieval Europe. *Guelph Ichthyology Reviews* 3, 57–85.

Hoffmann RC (1996) Economic development and aquatic ecosystems in medieval Europe. *American Historical Review* **101**(3), 631–668. doi:10.2307/2169418

Hoffmann RC (2005) A brief history of aquatic resource use in medieval Europe. *Helgoland Marine Research* **59**(1), 22–30. doi:10.1007/s10152-004-0203-5

Hoffmann RC (2023) *The Catch: An Environmental History of Medieval European Fisheries*. Cambridge University Press, Cambridge, UK.

Hofmeister E (2016) From carp to rainbow trout: freshwater fish production in Denmark. In *Historical Aquaculture in Northern Europe*. (Eds M Bonow, H Olsen and I Svanberg) pp. 77–88. Södertörn University, Södertörn, Sweden.

Homemade Boilies, UK (2023) *The Art of Homemade Boilies: A Comprehensive Guide to Crafting Perfect Carp Bait*. Gumroad.com, UK.

Hopf J, Davis S, Graham K, Stratford D, Durr PA (2024) Linking demographic and habitat suitability modelling identifies the environmental determinants of successfully controlling invasive common carp (*Cyprinus carpio*) in south-eastern Australia. *Biological Invasions* **26**(5), 651–670. doi:10.1007/s10530-023-03198-x

Horn C (2025) Invasive pest carp a popular part of traditional Czech-Australian Christmas. ABC Australia.

Howes GJ (1991) Systematics and biogeography: an overview. In *Cyprinid Fishes: Systematics, Biology and Exploitation*. (Eds IJ Winfield, JS Nelson) pp. 1–33. Springer, Dordrecht.

Howitt W (1858) *Land, Labour and Gold, Or, Two Years in Victoria: With Visits to Sydney and Van Diemen's Land*, Longman, Brown, Green, Longmans, and Roberts, Boston.

Hufthammer AK, Moe D (2016) Fishponds and aquaculture in historical times in Norway. In *Historical Aquaculture in Northern Europe*. (Eds M Bonow, H Olsen and I Svanberg) pp. 121–138. Södertörn University, Södertörn, Sweden.

Hughes R (1996) *The Fatal Shore: A History of the Transportation of Convicts to Australia, 1787–1868*. Harvill, London.

Hume DJ, Pribble HJ (1980) Annual Report. *Carp Program Report* 1980, No. 5. Arthur Rylah Institute for Environmental Research, Melbourne.

Hume DJ, Fletcher AR, Morison AK (1981) Annual Report. *Carp Program Report* 1980/1981, No. 8. Arthur Rylah Institute for Environmental Research, Melbourne.

Hume DJ, Fletcher AR, Morison AK (1983) Final Report. *Carp Program Report No. 10*. Arthur Rylah Institute for Environmental Research, Melbourne.

Humphries P (2023) *The Life and Times of the Murray Cod*. CSIRO Publishing, Melbourne.

Humphries P, Walker K (Ed) (2013) *Ecology of Australian Freshwater Fishes*. CSIRO Publishing, Melbourne.

Humphries P, Serafini LG, King AJ (2002) River regulation and fish larvae: variation through space and time. *Freshwater Biology* **47**(7), 1307–1331. doi:10.1046/j.1365-2427.2002.00871.x

Humphries P, Brown P, Douglas J, Pickworth A, Strongman R, *et al.* (2008) Flow-related patterns in abundance and composition of the fish fauna of a degraded Australian lowland river. *Freshwater Biology* **53**(4), 789–813. doi:10.1111/j.1365-2427.2007.01904.x

Hunter DA, Smith MJ, Scroggie MP, Gilligan D (2011) Experimental examination of the potential for three introduced fish species to prey on tadpoles of the endangered Booroolong frog, *Litoria booroolongensis*. *Journal of Herpetology* **45**, 181–185. doi:10.1670/10-010.1

Huser BJ, Bajer PG, Kittelson S, Christenson S, Menken K (2022) Changes to water quality and sediment phosphorus forms in a shallow, eutrophic lake after removal of common carp (*Cyprinus carpio*). *Inland Waters* **12**(1), 33–46. doi:10.1080/20442041.2020.1850096

Inland Fisheries Commission (1975) Inland Fisheries Commission Report for Year ending 30 June 1975. Hobart.

Inland Fisheries Commission (1980) Inland Fisheries Commission Report for Year ending 30 June 1980. Hobart.

Inland Fisheries Commission (1995) Inland fisheries commission newsletter. *Inland Fisheries Commission Newsletter* **24**, 12.

Inland Fisheries Service (2004) Inland Fisheries Service, Carp Management Program Report: Lakes Crescent and Sorell 1995–June 2004. Hobart.

Inland Fisheries Service (2010) Inland Fisheries Service, Carp Management Program Annual Report: 2009–2010. Hobart.

Inland Fisheries Service (2011) Inland Fisheries Service, Carp Management Program Annual Report: 2010–11. Hobart.

Inland Fisheries Service (2016) Inland Fisheries Service, Carp Management Program Quarterly Report: October to December 2016. Hobart.

Inland Fisheries Service (2023) Inland Fisheries Service, Carp Management Program Annual Report: 2022–23. Hobart.

Jackson L (2013) *Just for the Record: the Quest for Two-Tone.* Angling Publications Limited, Sheffield.

Jeffries S (1997) Alexander von Humboldt and Ferdinand von Mueller's argument for the scientific botanic garden. *Historical Records of Australian Science* **11**(3), 301–310. doi:10.1071/hr9971130301

Jenkins JT (1946) *The Fishes of the British Isles, Both Fresh Water and Salt*. F. Warne & Company, London.

Joehnk KD, Graham K, Sengupta A, Chen Y, Aryal SK, *et al.* (2020) The role of water temperature modelling in the development of a release strategy for cyprinid herpesvirus 3 (CyHV-3) for common carp control in south-eastern australia. *Water (Switzerland)* **12**(11), 1–25. doi:10.3390/w12113217

Johnston RM (1882) General and critical observations on the fishes of Tasmania; with a classified catalogue of all the known species. *Papers and Proceedings of the Royal Society of Tasmania* 1882, 53–144.

Jones H (1995) A fisherman's thoughts. In *Proceedings of the National Carp Summit*, 1995 Adelaide, pp. 26–29. Murray–Darling Association.

Jones MJ, Stuart IG (2007) Movements and habitat use of common carp (*Cyprinus carpio*) and Murray cod (*Maccullochella peelii peelii*) juveniles in a large lowland Australian river. *Ecology of Freshwater Fish* **16**(2), 210–220. doi:10.1111/j.1600-0633.2006.00213.x

Jones MJ, Stuart IG (2009) Lateral movement of common carp (*Cyprinus carpio* L.) in a large lowland river and floodplain. *Ecology of Freshwater Fish* **18**(1), 72–82. doi:10.1111/j.1600-0633.2008.00324.x

Kanitskiy S (1983) Structure of the spawning stock and spawning features of the Amur carp, *Cyprinus carpio haematopterus*, in the Barguzin river drainage. *Journal of Ichthyology* **33**, 189–193.

Kasumyan A, Kuzishchin K, Gruzdeva M (2024) Evaluation of the effectiveness of food chemical attractants for wild common carp, *Cyprinus carpio* (Cyprinidae) under conditions of natural water body. *Journal of Ichthyology* **64**(4), 689–704. doi:10.1134/S0032945224700279

Katano O, Hakoyama H, Matsuzaki SiS (2015) Japanese inland fisheries and aquaculture: status and trends. In *Freshwater Fisheries Ecology*. (Ed. JF Craig) pp. 231–240. John Wiley & Sons, Wiley Online.

Keller RP (2014) Ecological separation without hydraulic separation: engineering solutions to control invasive common carp in Australian rivers. In *Invasive Species in a Globalized World: Ecological, Social, & Legal Pespectives on Policy.* (Eds RP Keller, MW Cadotte and G Sandiford) p. 416. University of Chicago Press, Chicago.

Khan TA (2003) Dietary studies on exotic carp (*Cyprinus carpio* L.) from two lakes of western Victoria, Australia. *Aquatic Sciences* **65**(3), 272–286. doi:10.1007/s00027-003-0658-5

Khan TA, Wilson ME, Khan MT (2003) Evidence for invasive carp mediated trophic cascade in shallow lakes of western Victoria, Australia. *Hydrobiologia* **506**, 465–472. doi:10.1023/B:HYDR.0000008558.48008.63

King A (1995) *The Effects of Carp on Aquatic Ecosystems: A Literature Review.* A Report prepared for the Environmental Protection Authority NSW, Murray region; November 1995. Charles Sturt University, Wagga Wagga, NSW.

King A, Robertson A, Healey M (1997) Experimental manipulations of the biomass of introduced carp (*Cyprinus carpio*) in billabongs. I. Impacts on water-column properties. *Marine and Freshwater Research* **48**(5), 435–443. doi:10.1071/MF97031

King AJ (2005) Ontogenetic dietary shifts of fishes in an Australian floodplain river. *Marine and Freshwater Research* **56**(2), 215–225. doi:10.1071/MF04117

King AJ, Humphries P, Lake PS (2003) Fish recruitment on floodplains: the roles of patterns of flooding and life history characteristics. *Canadian Journal of Fisheries and Aquatic Sciences* **60**(7), 773–786. doi:10.1139/f03-057

King AJ, Humphries P, McCasker NG (2013) Reproduction and early life history. In *Ecology of Australian Freshwater Fishes.* (Eds P Humphries, KF Walker) pp. 159–192. CSIRO Publishing, Melbourne.

King DR, Hunt GS (1967) Effect of carp on vegetation in a Lake Erie marsh. *Journal of Wildlife Management* **31**(1), 181–188. doi:10.2307/3798375

Kirpitchnikov VS (1999) *Genetics and Breeding of Common Carp.* INRA Editions, Paris.

Kodama Koi Farm (2024) *How Much Do Koi Fish Cost? Koi Pricing Guide for USA.* Available: <https://www.kodamakoifarm.com/how-much-do-koi-fish-cost/#:~:text=Kohaku%20is%20generally%20the%20most,quality%20koi%2C%20hurts%20even%20more!>

Koehn J, Brumley AR, Gehrke PC (2000) *Managing the Impacts of Carp.* Bureau of Rural Sciences, Canberra.

Koehn J, Brown G, Clunie P, Menkhorst P, Woodford L, *et al.* (2022) *People Passion Science: Celebrating 50 Years of the Arthur Rylah Institute for Environmental Research.* Arthur Rylah Institute for Environmental Research, Melbourne.

Koehn JD (2004) Carp (*Cyprinus carpio*) as a powerful invader in Australian waterways. *Freshwater Biology* **49**(7), 882–894. doi:10.1111/j.1365-2427.2004.01232.x

Koehn JD, Nicol SJ (1998) Habitat and movement requirements of fish. In *Proceedings of the 1996 Riverine Environment Forum*, Brisbane. (Eds RJ Banens, R Lehane) pp. 1–6. Murray-Darling Basin Commission, Canberra.

Koehn JD, Nicol SJ (2014) Comparative habitat use by large riverine fishes. *Marine and Freshwater Research* **65**(2), 164–174. doi:10.1071/MF13011

Koehn JD, Nicol SJ (2016) Comparative movements of four large fish species in a lowland river. *Journal of Fish Biology* **88**(4), 1350–1368. doi:10.1111/jfb.12884

Koehn JD, Todd CR, Zampatti BP, Stuart IG, Conallin A, *et al.* (2018) Using a population model to inform the management of river flows and invasive carp (*Cyprinus carpio*). *Environmental Management* **61**(3), 432–442. doi:10.1007/s00267-017-0855-y

Koehn JD, Raymond SM, Stuart I, Todd CR, Balcombe SR, *et al.* (2020) A compendium of ecological knowledge for restoration of freshwater fishes in Australia's Murray–Darling Basin. *Marine and Freshwater Research* **71**(11), 1391–1463. doi:10.1071/MF20127

Kohlmann K (2015) The natural history of common carp and common carp genetics. In *Biology and Ecology of Carp.* (Eds C Pietsch, P Hirsch) pp. 3–26. Taylor & Francis, Boca Raton.

Kohlmann K, Kersten P (2013) Deeper insight into the origin and spread of European common carp (*Cyprinus carpio carpio*) based on mitochondrial D-loop sequence polymorphisms. *Aquaculture* **376–379**, 97–104. doi:10.1016/j.aquaculture.2012.11.006

Kohlmann K, Gross R, Murakaeva A, Kersten P (2003) Genetic variability and structure of common carp (*Cyprinus carpio*) populations throughout the distribution range inferred from allozyme, microsatellite and mitochondrial DNA markers. *Aquatic Living Resources* **16**(5), 421–431. doi:10.1016/S0990-7440(03)00082-2

Koi Society Australia (2024) *Koi Varieties* [online]. Available: <https://ksakoi.com/home/?page_id=622> (accessed 1 November 2024).

Komiyama T, Kobayashi H, Tateno Y, Inoko H, Gojobori T, *et al.* (2009) An evolutionary origin and selection process of goldfish. *Gene* **430**(1–2), 5–11. doi:10.1016/j.gene.2008.10.019

Kopf RK, Boutier M, Finlayson CM, Hodges K, Humphries P, *et al.* (2019a) Biocontrol in Australia: can a carp herpesvirus (CyHV-3) deliver safe and effective ecological restoration? *Biological Invasions* **21**(6), 1857–1870. doi:10.1007/s10530-019-01967-1.

Kopf RK, Humphries P, Bond NR, Sims NC, Watts RJ, *et al.* (2019b) Macroecology of fish community biomass–size structure: effects of invasive species and river regulation. *Canadian Journal of Fisheries and Aquatic Sciences* **76**(1), 109–122. doi:10.1139/cjfas-2017-0544

Koshida S (1931) Fish farming as side business for farmers. Niigata Farmers Association, Niigata, Japan.

Kottelat M, Britz R, Hui TH, Witte K-E (2006) Paedocypris, a new genus of south-east Asian cyprinid fish with a remarkable sexual dimorphism, comprises the world's smallest vertebrate. *Proceedings of the Royal Society B: Biological Sciences* **273**(1589), 895–899. doi:10.1098/rspb.2005.3419

Kulhanek SA, Ricciardi A, Leung B (2011) Is invasion history a useful tool for predicting the impacts of the world's worst aquatic invasive species? *Ecological Applications* **21**(1), 189–202. doi:10.1890/09-1452.1

Lake JS (1959) The freshwater fishes of New South Wales. NSW State Fisheries. *Research Bulletin* **5**, 1–20.

Lake JS (1967) *Freshwater Fish of the Murray–Darling River System: The Native and Introduced Fish Species*. Chief Secretary's Department, Sydney.

Lake JS (1978) *Australian Freshwater Fishes: An Illustrated Field Guide*. Thomas Nelson (Australia), Melbourne.

Langdon J (2002) *Horses, Oxen and Technological Innovation: The Use of Draught Animals in English Farming from 1066–1500*. Cambridge University Press, Cambridge, UK.

Lapin G (2020) *Fly Fishing for Urban Carp*. Available: <https://www.manictackleproject.com/blogs/manic-fly-fishing-blog/team-tuesday-urban-carp> (accessed 22 December 2024).

Lawrence S, Davies P (2014) The sludge question: the regulation of mine tailings in nineteenth-century Victoria. *Environment and History* **20**(3), 385–410. doi:10.3197/096734014X14031694156448

Lawrence S, Moon J, Davies P (2018) Gold rush environmental change and its potential impact on Aboriginal archaeological sites in Victoria. In *Excavations, Surveys and Heritage Management in Victoria*. (Eds C Spry, E Foley, D Frankel and S Lawrence) pp. 47–52. La Trobe University, Melbourne.

Lear L (1998) *Rachel Carson: Witness for Nature*. Macmillan, New York.

Lechelt JD, Bajer PG (2016) Elucidating the mechanism underlying the productivity-recruitment hypothesis in the invasive common carp. *Aquatic Invasions* **11**(4), 469–482. doi:10.3391/ai.2016.11.4.11

Lefever R (1993) The continuing saga of the butterfly koi. *Tropical Fish Hobbyist* **41**, 94–102.

Lenders H, Chamuleau T, Hendriks A, Lauwerier R, Leuven R, *et al.* (2016) Historical rise of waterpower initiated the collapse of salmon stocks. *Scientific Reports* **6**, 29269. doi:10.1038/srep29269

Levi HW (1952) Evaluation of wildlife importations. *Scientific Monthly* **74**(6), 315–322.

Lim H, Sorensen PW (2012) Common carp implanted with prostaglandin F 2α release a sex pheromone complex that attracts conspecific males in both the laboratory and field. *Journal of Chemical Ecology* **38**(2), 127–134. doi:10.1007/s10886-012-0062-5

Lin CH, Landos MA (2019) *National Carp Control Plan: Biosecurity Strategy for the Koi (*Cyprinus carpio*) Industry*. Fisheries Research and Development Corporation, Canberra.

Linhart O, Kudo S, Billard R, Slechta V, Mikodina EV (1995) Morphology, composition and fertilization of carp eggs: a review. *Aquaculture* **129**(1–4), 75–93. doi:10.1016/0044-8486(94)00230-L

Lintermans M (2023) *Fishes of the Murray–Darling Basin*. Australian River Restoration Centre, Canberra.

Locker A (2014) The social history of coarse angling in England AD 1750–1950. *Anthropozoologica* **49**(1), 99–107. doi:10.5252/az2014n1a07

Lougheed VL, Crosbie B, Chow-Fraser P (1998) Prediction on the effect of common carp exclusion on water quality, zooplankton, and submergent macrophytes in a Great Lakes wetland. *Canadian Journal of Fisheries and Aquatic Sciences* **55**(5), 1189–1197. doi:10.1139/f97-315

Low T (2001) *Feral Future*. Penguin Books Australia, Melbourne.

Lowerson J (1983) Izaak Walton: father of a dream. *History Today* **33**(12), 28–32.

Lucas AR (2005) Industrial milling in the ancient and medieval worlds: a survey of the evidence for an industrial revolution in medieval Europe. *Technology and Culture* **46**(1), 1–30. doi:10.1353/tech.2005.0026

Luo M, Lu G, Yin H, Wang L, Atuganile M, *et al.* (2021) Fish pigmentation and coloration: molecular mechanisms and aquaculture perspectives. *Reviews in Aquaculture* **13**(4), 2395–2412. doi:10.1111/raq.12583

Lyon JP, O'Connor JP (2008) Smoke on the water: can riverine fish populations recover following a catastrophic fire-related sediment slug? *Austral Ecology* **33**(6), 794–806. doi:10.1111/j.1442-9993.2008.01851.x

Mabuchi K, Song H (2014) The complete mitochondrial genome of the Japanese ornamental koi carp (*Cyprinus carpio*) and its implication for the history of koi. *Mitochondrial DNA* **25**(1), 35–36. doi:10.3109/19401736.2013.779261

Mabuchi K, Senou H, Suzuki T, Nishida M (2005) Discovery of an ancient lineage of *Cyprinus carpio* from Lake Biwa, central Japan, based on mtDNA sequence data, with reference to possible multiple origins of koi. *Journal of Fish Biology* **66**(6), 1516–1528. doi:10.1111/j.0022-1112.2005.00676.x

Mabuchi K, Miya M, Senou H, Suzuki T, Nishida M (2006) Complete mitochondrial DNA sequence of the Lake Biwa wild strain of common carp (*Cyprinus carpio* L.): further evidence for an ancient origin. *Aquaculture* **257**(1–4), 68–77. doi:10.1016/j.aquaculture.2006.03.040

Maccarinelli A (2020) *The Social and Economic Role of Freshwater Fish in Medieval England: A Zooarchaeological Approach*. PhD thesis. University of Sheffield, Sheffield.

Macdonald A, Wisniewski C (2003) The use of biotelemetry in controlling the common carp (*Cyprinus carpio*) in Lakes Crescent and Sorell. Inland Fisheries Service, Hobart.

Macdonald J, McNeil DG, Crook DA (2010) *Identification of Carp Recruitment Hotspots in the Lachlan River using Otolith Chemistry*. Invasive Animals Cooperative Research Centre, Canberra.

Maceda-Veiga A, López R, Green AJ (2017) Dramatic impact of alien carp *Cyprinus carpio* on globally threatened diving ducks and other waterbirds in Mediterranean shallow lakes. *Biological Conservation* **212**(Part A), 74–85. doi:10.1016/j.biocon.2017.06.002

Mahmud R, Purser J, Patil JG (2020) A novel testicular degenerative condition in a wild population of the common carp *Cyprinus carpio* (L). *Journal of Fish Diseases* **43**(9), 1065–1076. doi:10.1111/jfd.13216

Maitland PS, Linsell K (2006) *Guide to Freshwater Fish of Britain and Europe*. Philip's, London.

Maiztegui T, Baigún CRM, García de Souza JR, Weyl OL, Colautti DC (2019) Population responses of common carp *Cyprinus carpio* to floods and droughts in the Pampean wetlands of South America. *NeoBiota* **48**, 25–54. doi:10.3897/neobiota.48.34850

Malcolm S (1971) The status of the common or European carp, *Cyprinus carpio* in Victoria. BSc (Hons) thesis, Monash University, Melbourne.

Mallen-Cooper M, Stuart I, Hides-Pearson F, Harris J (1995) Fish Migration in the Murray River and Assessment of the Torrumbarry Fishway. Final report to the Murray–Darling Basin Commission. *Natural Resources Management Strategy Project N002*.

Mankad A, Zhang A, Carter L, Curnock M (2022) A path analysis of carp biocontrol: effect of attitudes, norms, and emotion on acceptance. *Biological Invasions* **24**(3), 709–723. doi:10.1007/s10530-021-02679-1

Marsh GP (1864) *Man and Nature; Or Physical Geography as Modified by Human Action*. Sampson Low, Son & Marston, London.

Marshall JC, Blessing JJ, Clifford SE, Hodges KM, Negus PM, *et al.* (2019) Ecological impacts of invasive carp in Australian dryland rivers. *Aquatic Conservation Marine and Freshwater Ecosystems* **29**(11), 1870–1889. doi:10.1002/aqc.3206

Matsuzaki SS, Usio N, Takamura N, Washitani I (2009) Contrasting impacts of invasive engineers on freshwater ecosystems: an experiment and meta-analysis. *Oecologia* **158**(4), 673–686. doi:10.1007/s00442-008-1180-1

Mazumder D, Johansen M, Saintilan N, Iles J, Kobayashi T, *et al.* (2012) Trophic shifts involving native and exotic fish during hydrologic recession in floodplain wetlands. *Wetlands* **32**, 267–275.

McColl KA, Sunarto A (2020) Biocontrol of the common carp (*Cyprinus carpio*) in Australia: a review and future directions. *Fishes* **5**(2), 17. doi:10.3390/fishes5020017

McColl KA, Cooke BD, Sunarto A (2014) Viral biocontrol of invasive vertebrates: lessons from the past applied to cyprinid herpesvirus-3 and carp (*Cyprinus carpio*) control in Australia. *Biological Control* **72**, 109–117. doi:10.1016/j.biocontrol.2014.02.014

McColl KA, Sunarto A, Holmes EC (2016) Cyprinid herpesvirus 3 and its evolutionary future as a biological control agent for carp in Australia. *Virology Journal* **13**(1), 1–4. doi:10.1186/s12985-016-0666-4

McColl KA, Sunarto A, Slater J, Bell K, Asmus M, *et al.* (2017) Cyprinid herpesvirus 3 as a potential biological control agent for carp (*Cyprinus carpio*) in Australia: susceptibility of non-target species. *Journal of Fish Diseases* **40**(9), 1141–1153. doi:10.1111/jfd.12591

McCoy F (1862) Acclimatisation, its nature and applicability to Victoria. *First Annual Report of the Acclimatisation Society of Victoria: with the Addresses Delivered at the Annual Meeting of the Society held November 24th, 1862, at the Mechanics Institute, Melbourne.* Acclimatisation Society of Victoria, Melbourne.

McCulloch AR (1929) A check-list of the fishes recorded from Australia. Part I. *Australian Museum Memoir* **5**, 1–144.

McDonnell J (1981) *Inland Fisheries in Medieval Yorkshire, 1066–1300.* Borthwick Publications, York.

McDonough B (2023) The koi carp fish in Japan: what is their significance? *Interac: Enrich through Education* [online]. Available: <https://interacnetwork.com/what-is-the-significance-of-the-koi-carp-in-japan/> (accessed 7 February 2024).

McGinness HM, Paton A, Gawne B, King AJ, Kopf RK, *et al.* (2019) Effects of fish kills on fish consumers and other water-dependent fauna: exploring the potential effect of mass mortality of carp in Australia. *Marine and Freshwater Research* **71**, 156–169. doi:10.1071/MF19035

McGowan B (2001) Mullock heaps and tailing mounds: environmental effects of alluvial goldmining. In *Gold: Forgotten Histories and Lost Objects of Australia.* (Eds I McCalman, A Cook and A Reeves) pp. 85–101. Cambridge University Press, Melbourne.

McLeod R (2004) *Counting the Cost: Impact of Invasive Animals in Australia, 2004.* Cooperative Research Centre for Pest Animal Control, Canberra.

McNeil DG, Closs GP (2007) Behavioural responses of a south-east Australian floodplain fish community to gradual hypoxia. *Freshwater Biology* **52**(3), 412–420. doi:10.1111/j.1365-2427.2006.01705.x

Michel P, Oberdorff T (1995) Feeding habits of fourteen European freshwater fish species. *Cybium* **19**(1), 5–46.

Minard P (2013) Assembling acclimatization: Frederick McCoy, European ideas, Australian circumstances. *Historical Records of Australian Science* **24**(1), 1–14. doi:10.1071/HR12017

Minard P (2015) Salmonid acclimatisation in colonial Victoria: improvement, restoration and recreation 1858–1909. *Environment and History* **21**(2), 177–199. doi:10.3197/096734015X14267043141345

Minard P (2019) *All Things Harmless, Useful, and Ornamental: Environmental Transformation through Species Acclimatization, from Colonial Australia to the World.* UNC Press Books, Chapel Hill, NC.

Mintram KS, van Oosterhout C, Lighten J (2021) Genetic variation in resistance and high fecundity impede viral biocontrol of invasive fish. *Journal of Applied Ecology* **58**(1), 148–157. doi:10.1111/1365-2664.13762

Mitchel B (2019) Plastic fantastic: using artificial baits. *BadAngling* [online]. Available: <https://badangling.com/carp/plastic-fantastic-using-artificial-baits/> (accessed 5 October 2023).

Montague GF, Schooley JD, Scarnecchia DL, Snow RA (2023) Bowfishing shoot and release: high short-term mortality of nongame fishes and its management implications. *North American Journal of Fisheries Management* **43**(3), 962–983. doi:10.1002/nafm.10904

Mulley J, Shearer K (1980) Identification of natural 'Yanco' × 'Boolara' hybrids of the carp, *Cyprinus carpio. Marine and Freshwater Research* **31**(3), 409–411. doi:10.1071/MF9800409

Murdy E, Birdsong R, Musick J, Secor D (1998) Fishes of Chesapeake Bay. *Reviews in Fish Biology and Fisheries* **8**, 105–105.

Murray–Darling Basin Commission (2002) *Carp: Villains or Victims*. National Carp Task Force, Canberra.

Mutethya E, Yongo E (2021) A comprehensive review of invasion and ecological impacts of introduced common carp (*Cyprinus carpio*) in Lake Naivasha, Kenya. *Lakes & Reservoirs: Research & Management* **26**(4), e12386. doi:10.1111/lre.12386

Nakajima T (1994) Cyprinid fishes. In *The Natural History of Lake Biwa*. (Ed. T Nakajima) pp. 235–275. Yasaka Shobo, Tokyo.

Nakajima T, Hudson MJ, Uchiyama J, Makibayashi K, Zhang J (2019) Common carp aquaculture in neolithic China dates back 8,000 years. *Nature Ecology & Evolution* **3**(10), 1415–1418. doi:10.1038/s41559-019-0974-3

Nasir NA, Yesser AT, Al-Hamadany QH (2019) Effect of pH on the growth and survival of juvenile common carp (*Cyprinus carpio* L.). *Iraqi Journal of Science* 60(1B), 234–238. doi:10.24996/ijs.2019.60.2.5

National Carp Control Plan (2022a) *Epidemiology and Release Strategies*. Fisheries Research and Development Corporation, Canberra.

National Carp Control Plan (2022b) *The National Carp Control Plan*. Fisheries Research and Development Corporation, Canberra.

National Carp Control Plan (2022c) *Potential Socio-economic Impacts of Carp Biocontrol*. Fisheries Research and Development Corporation, Canberra.

Nelson JS, Grande TC, Wilson MV (2016) *Fishes of the World*. John Wiley & Sons, Hoboken, NJ.

Nicol SJ, Lieschke JA, Lyon JP, Koehn JD (2004) Observations on the distribution and abundance of carp and native fish, and their responses to a habitat restoration trial in the Murray River, Australia. *New Zealand Journal of Marine and Freshwater Research* **38**(3), 541–551. doi:10.1080/00288330.2004.9517259

Niigata Prefecture Tourism Association (2024) *Ojiya Nishikigoi no Sato: A One-of-a-Kind Up-Close Encounter with Colorful 'Nishikigoi' Carp in a Serene Japanese Garden* [online]. Available: <https://enjoyniigata.com/en/spot/6096#:~:text=Nishikigoi%20no%20Sato%20is%20a,specimens%20from%2015%20different%20varieties> (accessed 1 November 2024).

Noakes DLG (1995) Carp: reviled and revered. *Guelph Ichthyology Reviews* **3**, preface.

Norris A (2011) Guidelines for planning carp fishing competitions. *PestSmart Toolkit Publication*. Invasive Animals Cooperative Research Centre, Canberra.

Norris A, Chilcott K, Hutchison M, Stewart D (2011) *Carp Surveys of the Logan and Albert Rivers Catchment, 2006–2009*. Invasive Animals Cooperative Research Centre, Canberra.

Norris A, Chilcott K, Hutchison M (2013) The role of fishing competitions in pest fish management. PestSmart Toolkit Publication. Invasive Animals Cooperative Research Centre, Canberra.

Norris A, Hutchison M, Chilcott K, Stewart D (2014) *Effectiveness of Carp Removal Techniques: Options for Local Governments and Community Groups*. PestSmart Toolkit Publication. Invasive Animals Cooperative Research Centre, Canberra.

North Central CMA (2019) *Catch a Carp this Gone Fishing Day* [online]. Available: <https://www.nccma.vic.gov.au/media-events/media-releases/catch-carp-gone-fishing-day-0> (accessed 20 March 2024).

Northern Suburbs Fly Fishing Club (2024) *Carp: Freshwater Bonefish* [online]. Available: <https://www.flyfishing.org.au/index.php/78-nsffc/85-carp-freshwater-bonefish> (accessed 21 January 2024).

NSW Department of Primary Industries (nd) Brochure on releasing aquarium fish. Available at: <https://www.dpi.nsw.gov.au>.

NSW Department of Primary Industries (2024) *Angler Access* [online]. Available: <https://www.dpi.nsw.gov.au/fishing/recreational/resources/angler-access> (accessed 2 November 2023).

O'Brien N, Cormack L (2016) Carp bow fishing trial threatens native species, opponents say. *Sydney Morning Herald*, 9 January 2016.

O'Loughlin LS, Green PT (2017) Secondary invasion: when invasion success is contingent on other invaders altering the properties of recipient ecosystems. *Ecology and Evolution* 7(19), 7628–7637. doi:10.1002/ece3.3315

Opuszyński K, Lirski A, Myszkowski L, Wolnicki J (1989) Upper lethal and rearing temperatures for juvenile common carp, *Cyprinus carpio* L., and silver carp, *Hypophthalmichthys molitrix* (Valenciennes). *Aquaculture Research* **20**(3), 287–294. doi:10.1111/j.1365-2109.1989.tb00354.x

Osborne MA (2000) Acclimatizing the world: a history of the paradigmatic colonial science. *Osiris* **15**, 135–151. doi:10.1086/649323

Ott ME, Heisler N, Ultsch GR (1980) A re-evaluation of the relationship between temperature and the critical oxygen tension in freshwater fishes. *Comparative Biochemistry and Physiology Part A: Physiology* **67**(3), 337–340. doi:10.1016/S0300-9629(80)80005-3

Oyen F, Camps L, Bonga SW (1991) Effect of acid stress on the embryonic development of the common carp (*Cyprinus carpio*). *Aquatic Toxicology* **19**(1), 1–12. doi:10.1016/0166-445X(91)90024-4

Oyugi D, Cucherousset J, Baker D, Britton J (2012) Effects of temperature on the foraging and growth rate of juvenile common carp, *Cyprinus carpio*. *Journal of Thermal Biology* **37**(1), 89–94. doi:10.1016/j.jtherbio.2011.11.005

OzFish (2023) *Catch a Carp Day Wentworth Wharf* [online]. Available: <https://ozfish.org.au/catch-a-carp-day-wentworth-wharf-august-2023-rules/> (accessed 20 March 2024).

Panek FM (1987) Biology and ecology of carp. In *Carp in North America*. (Ed. EL Cooper) pp. 1–16. American Fisheries Society, Bethesda, MD.

Patil JG, Wisniewski C (2011) *Hypophysation: A Technique for Deployment of Odour Donor Fish for Control of Common Carp (Cyprinus carpio)*. Inland Fisheries Service, Hobart.

Patowary K (2019) Monet's pond: the pond where art comes to life. *Amusing Planet* [online]. Available: <https://www.amusingplanet.com/2019/08/monets-pond-pond-where-art-comes-to-life.html> (accessed 21 August 2024).

Pattist K (2015) How a koi became a dragon: the waterfall legend. *Koi Organisation International* [online]. Available: <https://koiorganisationinternational.org/blog-entry/how-koi-became-dragon-waterfall-legend> (accessed 10 May 2024).

Pattist K (2016) Genetics. *Koi Organisation International* [online]. Available: <https://koiorganisationinternational.org/blog-entry/genetics> (accessed 22 October 2024).

Pavia L (2023) 'Trying to get a bank spot is crazy as it's become so popular': how gen Z got hooked on urban fishing. *The Guardian*, 2 June 2023.

Pavlov DA, Smirnov SA, Chernyaev ZA (2017) Eugene Kornel Balon (1930–2013): in memory of our colleague. *Journal of Ichthyology* **57**(3), 484–489. doi:10.1134/S0032945217030018

Pearn J (2020) The Queensland acclimatisation society. *Queensland History Journal* **24**, 339–355.

Pearson J (2019) Fly fishing for 'sewer salmon' in the L.A. River. *Los Angeles Times*, 11 July 2019.

Pease B, Grinberg A (1995) *New South Wales Commercial Fisheries Statistics 1940 to 1992*. Fisheries Research Institute, NSW Fisheries.

Perelberg A, Ronen A, Hutoran M, Smith Y, Kotler M (2005) Protection of cultured *Cyprinus carpio* against a lethal viral disease by an attenuated virus vaccine. *Vaccine* **23**(26), 3396–3403. doi:10.1016/j.vaccine.2005.01.096

Pfennigwerth S (2013) New creatures made known: some animals' histories of the Baudin expedition. In *Discovery and Empire: The French in the South Seas*. (Ed. J West-Sooby) pp. 171–214. University of Adelaide Press, Adelaide.

Piczak M, Brooks J, Boston C, Doka S, Portiss R, *et al.* (2023) Spatial ecology of non-native common carp (*Cyprinus carpio*) in Lake Ontario with implications for management. *Aquatic Sciences* **85**(1), 20. doi:10.1007/s00027-022-00917-9

Pillar J (1976) Decline of the native fish species in the River Murray. *SAFIC* **1**, 19–24.

PIRSA (2022) *Bow and Arrow* [online]. Available: <https://pir.sa.gov.au/recreational_fishing/rules/gear_bait_and_berley/permitted_fishing_gear/bow_and_arrow> (accessed 2 November 2023).

Pokorova D, Vesely T, Piackova V, Reschova S, Hulova J (2005) Current knowledge on koi herpesvirus (KHV): a review. *Veterinarni Medicina* **50**(4), 139–147. doi:10.17221/5607-VETMED

Poole JR, Bajer PG (2019) A small native predator reduces reproductive success of a large invasive fish as revealed by whole-lake experiments. *PLOS One* **14**(4), e0214009. doi:10.1371/journal.pone.0214009

Popper AN, Carlson TJ (1998) Application of sound and other stimuli to control fish behavior. *Transactions of the American Fisheries Society* **127**(5), 673–707. doi:10.1577/1548-8659(1998)127<0673:AOSAOS>2.0.CO;2

Pounds NJG (2014) *An Economic History of Medieval Europe*. Routledge, London.

Powell JM (1989) *Watering the Garden State: Water, Land and Community in Victoria, 1834–1988*. Allen & Unwin, Sydney.

Pribble HJ (1980) Carp program annual report 1979/1980. Fisheries and Wildlife Division, Ministry for Conservation Victoria, Melbourne.

Ragnarsson-Stabo H (2015) Recreational fishing for carp: implications for management and growth of carp populations. In *Biology and Ecology of Carp*. (Eds C Pietsch, P Hirsch) pp. 282–300. CRC Press, Boca Raton.

Rainboth WJ (1996) *Fishes of the Cambodian Mekong. FAO Species Identification Field Guide for Fishery Purposes*. FAO, Rome.

Rajeshkumar S, Liu Y, Ma J, Duan HY, Li X (2017) Effects of exposure to multiple heavy metals on biochemical and histopathological alterations in common carp, *Cyprinus carpio* L. *Fish & Shellfish Immunology* **70**, 461–472. doi:10.1016/j.fsi.2017.08.013

Rayner TS, Creese RG (2006) A review of rotenone use for the control of non-indigenous fish in Australian fresh waters, and an attempted eradication of the noxious fish, *Phalloceros caudimaculatus*. *New Zealand Journal of Marine and Freshwater Research* **40**(3), 477–486. doi:10.1080/00288330.2006.9517437

Reid D, Harris J, Chapman DJ (1997) *NSW Inland Commercial Fishery Data Analysis*. Fisheries Research & Development Corporation, Canberra.

Reynolds LF (1976) Tagging important in River Murray fish study. *Australian Fisheries Newsletter* **35**(7), 4–6.

Reynolds LF (1983) Migration patterns of five fish species in the Murray–Darling River system. *Marine and Freshwater Research* **34**(6), 857–871. doi:10.1071/MF9830857

Rhodes JO (1999) *Heads and Tales: Recollections of a Fisheries and Wildlife Officer*. Australian Deer Research Foundation, Melbourne.

Richards MP, Karavanić I, Pettitt P, Miracle P (2015) Isotope and faunal evidence for high levels of freshwater fish consumption by Late Glacial humans at the Late Upper Palaeolithic site of Šandalja II, Istria, Croatia. *Journal of Archaeological Science* **61**, 204–212. doi:10.1016/j.jas.2015.06.008

Richardson WB, Wickham SA, Threlkeld ST (1990) Foodweb response to the experimental manipulation of a benthivore (*Cyprinus carpio*), zooplanktivore (*Menidia beryllina*) and benthic insects. *Archiv fur Hydrobiologie* **119**(2), 143–165. doi:10.1127/archiv-hydrobiol/119/1990/143

Roberts AL (1969) European carp in Gippsland. *Fur Feathers Fins* **121**, 14–16.

Roberts J, Sainty GR (1996) *Listening to the Lachlan*. Murray–Darling Basin Commission, Canberra.

Roberts J, Tilzey R (Eds) (1997) *Controlling Carp: Exploring the Options for Australia. Proceedings of a Workshop*, 22–24 October 1996, Albury, NSW. CSIRO Land and Water, Griffith, NSW.

Roberts J, Chick A, Oswald L, Thompson P (1995) Effect of carp, *Cyprinus carpio* L., an exotic benthivorous fish, on aquatic plants and water quality in experimental ponds. *Marine and Freshwater Research* **46**(8), 1171–1180. doi:10.1071/MF9951171

Roberts K (2023) Unlocking the secrets of fish coloration: how do fish get their colors? *Fishy Features* [online]. Available: <https://fishyfeatures.com/unlocking-the-secrets-of-fish-coloration-how-do-fish-get-their-colors/> (accessed 19 March 2024).

Robertson A, Healey M, King A (1997) Experimental manipulations of the biomass of introduced carp (*Cyprinus carpio*) in billabongs. II. Impacts on benthic properties and processes. *Marine and Freshwater Research* **48**(5), 445–454. doi:10.1071/MF97032

Rolls EC (1969) *They All Ran Wild: The Story of Pests on the Land in Australia*. Angus & Robertson, Sydney.

Roughley TC (1951) *Fish and Fisheries of Australia*. Angus & Robertson, Sydney.

Royal Thai Art International (2024) *Koi Fish* [online]. Available: <https://royalthaiart.com/painting-category/koi-fish-paintings/page/5/> (accessed 1 November 2024).

Rutherfurd ID, Kenyon C, Thoms M, Grove J, Turnbull J, *et al.* (2020) Human impacts on suspended sediment and turbidity in the River Murray, south-eastern Australia: multiple lines of evidence. *River Research and Applications* **36**(4), 522–541. doi:10.1002/rra.3566

Samsing F, Hopf J, Davis S, Wynne J, Durr P (2021) Will Australia's common carp (*Cyprinus carpio*) populations develop resistance to Cyprinid herpesvirus 3 (CyHV-3) if released as a biocontrol agent? Identification of pathways and knowledge gaps. *Biological Control* **157**, 104571. doi:10.1016/j.biocontrol.2021.104571

Sanders B, Morris N (2018) Aboriginal rangers net thousands of carp in new project to rid the Murray–Darling of the pest fish. *ABC News*.

Sapkale P, Singh R, Desai A (2011) Optimal water temperature and pH for development of eggs and growth of spawn of common carp (*Cyprinus carpio*). *Journal of Applied Animal Research* **39**(4), 339–345. doi:10.1080/09712119.2011.620269

Sarig S (1966) Synopsis of biological data on common carp *Cyprinus carpio* (Linnaeus) 1758 (Near East and Europe). FAO Fisheries Synopsis 31.2. FAO, Rome.

Scarnecchia DL, Schooley JD (2020) Bowfishing in the United States: history, status, ecological impact, and a need for management. *Transactions of the Kansas Academy of Science* **123**, 285–338. doi:10.1660/062.123.0301

Schilling HT, Butler GL, Cheshire KJ, Gilligan DM, Stocks JR, *et al.* (2024) Contribution of invasive carp (*Cyprinus carpio*) to fish biomass in rivers of the Murray–Darling Basin, Australia. *Biological Invasions* **26**(9), 1–17. doi:10.1007/s10530-024-03362-x

Scott TB, Glover CJM, Southcott RV (1974) *The Marine and Freshwater Fishes of South Australia*. A.B. James, Government Printer, Adelaide.

Scott WB, Crossman EJ (1973) Freshwater fishes of Canada. *Bulletin of the Fisheries Research Board of Canada* **184**, 1–966.

Secord J (2016) Introduced species, hybrid plants and animals, and transformed lands in the Hellenistic and Roman worlds. In *Routledge Handbook of Identity and the Environment in the Classical and Medieval Worlds*. (Eds RF Kennedy, M Jones-Lewis) pp. 210–229. Routledge, London and New York.

Selleck R (2001) Rocks in his head: importing European science to a nineteenth-century Australian university. *Paedagogica Historica* **37**(1), 153–173. doi:10.1080/0030923010370110

Shearer KD, Mulley JC (1978) The introduction and distribution of the carp, *Cyprinus carpio* Linnaeus, in Australia. *Australian Journal of Marine and Freshwater Research* **29**(5), 551–563. doi:10.1071/MF9780551

Sibbing FA (1982) Pharyngeal mastication and food transport in the carp (*Cyprinus carpio* L.): a cineradiographic and electromyographic study. *Journal of Morphology* **172**(2), 223–258. doi:10.1002/jmor.1051720208

Sibbing FA (1988) Specializations and limitations in the utilization of food resources by the carp, *Cyprinus carpio*: a study of oral food processing. *Environmental Biology of Fishes* **22**, 161–178. doi:10.1007/BF00005379

Sibbing FA, Osse JW, Terlouw A (1986) Food handling in the carp (*Cyprinus carpio*): its movement patterns, mechanisms and limitations. *Journal of Zoology* **210**(2), 161–203. doi:10.1111/j.1469-7998.1986.tb03629.x

Silva LG, Doyle KE, Duffy D, Humphries P, Horta A, *et al.* (2020) Mortality events resulting from Australia's catastrophic fires threaten aquatic biota. *Global Change Biology* **26**(10), 5345–5350. doi:10.1111/gcb.15282

Simonson MA, Weber MJ, McCombs A (2022) Hyperstability in electrofishing catch rates of common carp and bigmouth buffalo. *North American Journal of Fisheries Management* **42**(2), 425–437. doi:10.1002/nafm.10758

Sinclair P (2001) *The Murray: A River and Its People*. Melbourne University Press, Melbourne.

Sivakumaran KP, Brown P, Stoessel D, Giles A (2003) Maturation and reproductive biology of female wild carp, *Cyprinus carpio*, in Victoria, Australia. *Environmental Biology of Fishes* **68**(3), 321–332. doi:10.1023/A:1027381304091

Sloan JL, Cordo EB, Mensinger AF (2013) Acoustical conditioning and retention in the common carp (*Cyprinus carpio*). *Journal of Great Lakes Research* **39**(3), 507–512. doi:10.1016/j.jglr.2013.05.004

Smith-Root (2025) *Fish Guidance and Deterrence Barriers* [online]. Available: <https://www.smith-root.com/barriers/> (accessed 2 January 2025).

Smith BB (2004) Carp (*Cyprinus carpio* L.) spawning dynamics and early growth in the lower River Murray, South Australia. PhD thesis, University of Adelaide, Adelaide.

Smith BB (2005) The state of the art: a synopsis of information on common carp (*Cyprinus carpio*) in Australia. *SARDI Research Report Series No. 77.* South Australian Research and Development Institute, Adelaide.

Smith BB, Walker KF (2003) Validation of the ageing of 0+ carp (*Cyprinus carpio* L.). *Marine and Freshwater Research* **54**(8), 1005–1008. doi:10.1071/MF03010

Smith BB, Walker KF (2004a) Reproduction of common carp in South Australia, shown by young-of-the-year samples, gonadosomatic index and the histological staging of ovaries. *Transactions of the Royal Society of South Australia* **128**(2), 249–257.

Smith BB, Walker KF (2004b) Spawning dynamics of common carp in the River Murray, South Australia, shown by macroscopic and histological staging of gonads. *Journal of Fish Biology* **64**(2), 336–354. doi:10.1111/j.0022-1112.2004.00293.x

Smith O, Momber G, Bates R, Garwood P, Fitch S, *et al.* (2015) Sedimentary DNA from a submerged site reveals wheat in the British Isles 8000 years ago. *Science* **347**(6225), 998–1001. doi:10.1126/science.1261278

Sorensen PW, Rue MCP, Leese JM, Ghosal R, Lim H (2019) A blend of F Prostaglandins functions as an attractive sex pheromone in silver carp. *Fishes* **4**(2), 27. doi:10.3390/fishes4020027

Souza A, Argillier C, Blabolil P, Děd V, Jarić I, *et al.* (2022) Empirical evidence on the effects of climate on the viability of common carp (*Cyprinus carpio*) populations in European lakes. *Biological Invasions* **24**, 1–15. doi:10.1007/s10530-021-02710-5

Spates WH (2013) Shakespeare, Walton, and the 'Carp of Truth'. *ANQ: A Quarterly Journal of Short Articles, Notes and Reviews* **26**(1), 1–4. doi:10.1080/0895769X.2013.749152

Spear MJ, Walsh JR, Ricciardi A, Zanden MJV (2021) The invasion ecology of sleeper populations: prevalence, persistence, and abrupt shifts. *Bioscience* **71**(4), 357–369. doi:10.1093/biosci/biaa168

Stead D (1929) Introduction of the great carp *Cyprinus carpio* into waters of New South Wales. *Australian Zoologist* **6**, 100–102.

Stecyk J, Farrell A (2002) Cardiorespiratory responses of the common carp (*Cyprinus carpio*) to severe hypoxia at three acclimation temperatures. *Journal of Experimental Biology* **205**(Pt 6), 759–768. doi:10.1242/jeb.205.6.759

Steffens W (2008) *Der Karpfen, Cyprinus carpio.* Ziemsen Verlag, Wittenberg Lutherstadt.

Stillwell GT (1969) Morton Allport (1830–1878). *Australian Dictionary of Biography.* National Centre of Biography, Australian National University, Canberra.

Strahan R (1992) The origin and early history of the Royal Zoological Society of New South Wales. *Australian Zoologist* **28**, 6–10. doi:10.7882/AZ.1992.002

Stuart I, Fanson B, Lyon J, Stocks J, Brooks S, *et al.* (2021) Continental threat: how many common carp (*Cyprinus carpio*) are there in Australia? *Biological Conservation* **254**(4), 108942. doi:10.1016/j.biocon.2020.108942

Stuart IG, Conallin AJ (2018) Control of globally invasive common carp: an 11-year commercial trial of the Williams' cage. *North American Journal of Fisheries Management* **38**(5), 1160–1169. doi:10.1002/nafm.10221

Stuart I, Jones M (2002) Ecology and management of common carp in the Barmah-Millewa forest: Final Report of the Point Source Management of Carp Project to Agriculture Fisheries & Forestry Australia. Arthur Rylah Institute for Environmental Research, Melbourne.

Stuart IG, Jones M (2006a) Large, regulated forest floodplain is an ideal recruitment zone for non-native common carp (*Cyprinus carpio* L.). *Marine and Freshwater Research* **57**(3), 333–347. doi:10.1071/MF05035

Stuart IG, Jones M (2006b) Movement of common carp, *Cyprinus carpio*, in a regulated lowland Australian river: implications for management. *Fisheries Management and Ecology* **13**(4), 213–219. doi:10.1111/j.1365-2400.2006.00495.x

Stuart IG, Williams A, McKenzie J, Holt T (2006) Managing a migratory pest species: a selective trap for common carp. *North American Journal of Fisheries Management* **26**(4), 888–893. doi:10.1577/M05-205.1

Swee UB, McCrimmon HR (1966) Reproductive biology of the carp, *Cyprinus carpio* L., in Lake St Lawrence, Ontario. *Transactions of the American Fisheries Society* **95**(4), 372–380. doi:10.1577/1548-8659(1966)95[372:RBOTCC]2.0.CO;2

Swirepik J (1999) Physical disturbance of *Potamogeton tricarinatus* and sediment by carp (*Cyprinus carpio*) in experimental ponds. Masters thesis, University of Canberra, Canberra.

Tamadachi M (1994) *Cult of the Koi*. 2nd edn. TFH Publications, Neptune City, NJ.

Tamaki T, Aoki JT (1977) *Nishikigoi: Fancy Koi*. Tamaki Yogyoen, Hiroshima.

Taylor AH, Tracey SR, Hartmann K, Patil JG (2012) Exploiting seasonal habitat use of the common carp, *Cyprinus carpio*, in a lacustrine system for management and eradication. *Marine and Freshwater Research* **63**(7), 587–597. doi:10.1071/MF11252

Taylor C (2000) Medieval ornamental landscapes. *Landscapes* **1**(1), 38–55. doi:10.1179/lan.2000.1.1.38

Teem JL, Gutierrez JB (2014) Combining the Trojan Y chromosome and daughterless carp eradication strategies. *Biological Invasions* **16**, 1231–1240. doi:10.1007/s10530-013-0476-1

Teem JL, Gutierrez JB, Parshad RD (2014) A comparison of the Trojan Y chromosome and daughterless carp eradication strategies. *Biological Invasions* **16**, 1217–1230. doi:10.1007/s10530-013-0475-2

The Australian Koi Association Inc (2006) Joint submission to the Ornamental Fish Policy Working Group. Department of Agriculture, Fisheries and Forestry. Australian Koi Association, Sydney.

The Carp Society (2024) Available: <https://www.thecarpsociety.com/> (accessed 3 January 2024).

The Daily Telegraph (2017) Join the Catch a Carp competition at Eagle Vale pond. Macarthur Chronicle Campelltown.

The Riverlander (1974) Conference against carp. *The Riverlander*, September.

Thresher R (1997) Physical removal as an option for the control of feral carp populations. In *Controlling Carp: Exploring the Options for Australia. Proceedings of a Workshop*, 22–24 October 1996, Albury, NSW. (Eds J Roberts, R Tilzey) CSIRO Land and Water, Griffith, NSW.

Thresher R, Bax N. (2003) The science of producing daughterless technology; possibilities for population control using daughterless technology; maximising the impact of carp control. In *Proceedings of the National Carp Control Workshop*, 2003, pp. 19–24. Cooperative Research Centre for Pest Animal Control, Canberra.

Thresher RE (2008) Autocidal technology for the control of invasive fish. *Fisheries* **33**(3), 114–121. doi:10.1577/1548-8446-33.3.114

Thresher RE, Hayes K, Bax NJ, Teem J, Benfey TJ, *et al.* (2014) Genetic control of invasive fish: technological options and its role in integrated pest management. *Biological Invasions* **16**(6), 1201–1216. doi:10.1007/s10530-013-0477-0

Thresher RE, Allman J, Stremick-Thompson L (2018) Impacts of an invasive virus (CyHV-3) on established invasive populations of common carp (*Cyprinus carpio*) in North America. *Biological Invasions* **20**(395), 1703–1718. doi:10.1007/s10530-017-1655-2

Thwaites L, Cheshire D (2016) A pilot trial to evaluate novel carp monitoring techniques at wetland carp exclusion screens. A technical report to the Department of Environment, Water and Natural Resources, Adelaide.

Thwaites L, Schmarr D (2019) *Evaluating the Effectiveness of Carp Exclusion Screens at Wetland Inlets in the River Murray, South Australia*. South Australian Research and Development Institute (Aquatic Sciences), Adelaide.

Thwaites LA, Smith BB, Decelis M, Fleer D, Conallin A (2010) A novel push trap element to manage carp (*Cyprinus carpio* L.): a laboratory trial. *Marine and Freshwater Research* **61**(1), 42–48. doi:10.1071/MF09011

Todd CR, Koehn JD, Stuart IG, Wootton HF, Zampatti BP, *et al.* (2024) Modelling the response of common carp (*Cyprinus carpio*) to natural and managed flows using a stochastic population model. *Biological Invasions* **26**, 1437–1456.

Tonkin ZD, Humphries P, Pridmore PA (2006) Ontogeny of feeding in two native and one alien fish species from the Murray–Darling Basin, Australia. *Environmental Biology of Fishes* **76**(2/4), 303–315. doi:10.1007/s10641-006-9034-3

Trueman W (2011) *True Tales of the Trout Cod: River Histories of the Murray–Darling Basin*. Murray–Darling Basin Authority, Canberra.

Tucker J (2016) Carp, enough of the golden bones. It's time to give carp their due. *Gink and Gasoline* [online]. Available: <https://www.ginkandgasoline.com/carp/carp-enough-of-the-golden-bones/> (accessed 13 October 2016).

Turnpenny A, O'Keeffe N (2005) Screening for intake and outfalls: a best practice guide. Environment Agency Science Report SC030231.

Tyrrell I (2004) Acclimatisation and environmental renovation: Australian perspectives on George Perkins Marsh. *Environment and History* **10**(2), 153–167. doi:10.3197/0967340041159812

Van den Thillart G, van Waarde A (1985) Teleosts in hypoxia: aspects of anaerobic metabolism. *Molecular Physiology* **8**, 393–409.

Van Zen N (2023) Butterfly koi: origin, characteristics, how much does it cost? *Zen Koi Garden: In Depth Information About Koi Fish* [in Vietnamese; online]. Available: <https://zenkoifarm.vn/ca-koi-buom-gia-bao-nhieu-dac-diem-va-cach-cham-soc-nhu-the-nao/> (accessed 19 September 2024).

Victorian Fisheries Authority (2022) *Illegal Fishing Equipment* [online]. Available: <https://vfa.vic.gov.au/recreational-fishing/recreational-fishing-guide/fishing-equipment/illegal-fishing-equipment> (accessed 2 November 2023).

Vila-Gispert A, Moreno-Amich R (2002) Life-history patterns of 25 species from European freshwater fish communities. *Environmental Biology of Fishes* **65**(4), 387–400. doi:10.1023/A:1021181022360

Vilizzi L (1998) Age, growth and cohort composition of 0+ carp in the River Murray, Australia. *Journal of Fish Biology* **52**(5), 997–1013. doi:10.1111/j.1095-8649.1998.tb00599.x

Vilizzi L (2012) The common carp, *Cyprinus carpio*, in the Mediterranean region: origin, distribution, economic benefits, impacts and management. *Fisheries Management and Ecology* **19**(2), 93–110. doi:10.1111/j.1365-2400.2011.00823.x

Vilizzi L (2018) Age determination in common carp *Cyprinus carpio*: history, relative utility of ageing structures, precision and accuracy. *Reviews in Fish Biology and Fisheries* **28**(B), 461–484. doi:10.1007/s11160-018-9514-5

Vilizzi L, Copp G (2017) Global patterns and clines in the growth of common carp *Cyprinus carpio*. *Journal of Fish Biology* **91**(1), 3–40. doi:10.1111/jfb.13346

Vilizzi L, Walker KF (1999a) Age and growth of the common carp, *Cyprinus carpio*, in the River Murray, Australia: validation, consistency of age interpretation, and growth models. *Environmental Biology of Fishes* **54**(1), 77–106. doi:10.1023/A:1007485307308

Vilizzi L, Walker KF (1999b) The onset of the juvenile period in carp, *Cyprinus carpio*: a literature survey. *Environmental Biology of Fishes* **56**(1), 93–102. doi:10.1023/A:1007552601704

Vilizzi L, Thwaites LA, Smith BB, Nicol JM, Madden CP (2014) Ecological effects of common carp (*Cyprinus carpio*) in a semi-arid floodplain wetland. *Marine and Freshwater Research* **65**(9), 802–817. doi:10.1071/MF13163

Vilizzi L, Tarkan AS, Copp G (2015) Experimental evidence from causal criteria analysis for the effects of common carp *Cyprinus carpio* on freshwater ecosystems: a global perspective. *Reviews in Fisheries Science & Aquaculture* **23**(3), 253–290. doi:10.1080/23308249.2015.1051214

Visit Canberra (2024) *Bredbo Australia Day Carp Out Competition* [online]. Available: <https://visitcanberra.com.au/events/658e2096eb0a1556129dda8e/bredbo-australia-day-carp-out-competition> (accessed 22 January 2024).

Vitousek PM, D'antonio CM, Loope LL, Rejmanek M, Westbrooks R (1997) Introduced species: a significant component of human-caused global change. *New Zealand Journal of Ecology* **21**(1), 1–16.

Walker R, Donkers P (2011) An examination of the selectivity of fishing equipment in relation to controlling carp (*Cyprinus carpio*) in Lakes Sorell and Crescent. Technical Report. Inland Fisheries Service, Hobart.

Walton I (1909) *The Compleat Angler*. Cassell & Co, London.

Wang J-T, Li J-T, Zhang X-F, Sun X-W (2012) Transcriptome analysis reveals the time of the fourth round of genome duplication in common carp (*Cyprinus carpio*). *BMC Genomics* **13**, 1–10. doi:10.1186/1471-2164-13-96

Wang Q, Sun C, Liu L (2024) Ecological impacts of common carp invasions: a global perspective. *International Journal of Aquaculture* **14**(4), 174–183. doi:10.5376/ija.2024.14.0018

Weatherley A (1974) Introduced freshwater fish. In *Biogeography and Ecology in Tasmania.* (Ed. WD Williams) pp. 141–170. Junk, The Hague.

Weatherley A, Lake J (1967) Introduced fish species in Australian inland waters. In *Australian Inland Waters and Their Fauna: Eleven Studies.* (Ed. A Weatherley) pp. 217–239. ANU Press, Canberra.

Weber MJ, Brown ML (2009) Effects of common carp on aquatic ecosystems 80 years after 'carp as a dominant': ecological insights for fisheries management. *Reviews in Fisheries Science* **17**(4), 524–537. doi:10.1080/10641260903189243

Weber MJ, Brown ML (2012) Maternal effects of common carp on egg quantity and quality. *Journal of Freshwater Ecology* **27**(3), 409–417. doi:10.1080/02705060.2012.666890

Weber MJ, Hennen MJ, Brown ML (2011) Simulated population responses of common carp to commercial exploitation. *North American Journal of Fisheries Management* **31**(2), 269–279. doi:10.1080/02755947.2011.574923

Weber MJ, Brown ML, Wahl DH, Shoup DE (2015) Metabolic theory explains latitudinal variation in common carp populations and predicts responses to climate change. *Ecosphere* **6**(4), 1–16. doi:10.1890/ES14-00435.1

Webster C (2021) Record turnout for Riverland carp fishing frenzy. *InDaily*, 16 March.

Wharton J (1971) European carp in Victoria. *Fur, Feathers and Fins* **130**, 3–11.

Wharton J (1979) *Impact of Exotic Animals, Especially European Carp Cyprinus carpio, on Native Fauna.* Fisheries and Wildlife Division, Ministry for Conservation Victoria, Melbourne.

White LT (1962) *Medieval Technology and Social Change.* Clarendon Press, Oxford.

Whitelaw I (2015) *The History of Fly Fishing in Fifty Flies.* White Lion Publishing, London.

Whiterod NR, Walker KF (2006) Will rising salinity in the Murray–Darling Basin affect common carp (*Cyprinus carpio* L.)? *Marine and Freshwater Research* **57**(8), 817–823. doi:10.1071/MF06021

Whitley G (1951) Introduced fishes. II. *Australian Museum Magazine* **10**, 234–238.

Whitlock D (2021) Dave Whitlock's 'Stalking the Golden Ghost': the lowly carp may be our smartest and most challenging freshwater gamefish. *Fly Fisherman* [online]. Available: <https://www.flyfisherman.com/editorial/dave-whitlocks-stalking-the-golden-ghost/453894> (accessed 8 November 2024).

Wieser W (1991) Physiological energetics and ecophysiology. In *Cyprinid Fishes: Systematics, Biology and Exploitation.* (Eds C Pietsch, P Hirsch) pp. 426–455. Springer, Canada.

Williamson M, Fitter A (1996) The varying success of invaders. *Ecology* **77**(6), 1661–1666. doi:10.2307/2265769.

Wilson E (1857) On the Murray River cod, with particulars of experiments instituted for introducing this fish into the River Yarra-Yarra. *Transactions of the Philosophical Institute of Victoria* **2**, 23–34.

Wilson GR (1998) Opportunities and constraints on the commercial use of carp in NSW. Industry Investment Brief prepared for the NSW Department of State and Regional Development in Association with Australia's Holiday Coast Development Board.

Wiltshire RJ, Barclay N, Kane HE, Kenny FR, Machin B, *et al.* (1962a) Report of the State Development Committee on the introduction of European carp into Victorian waters. A. C. Brooks, Government Printer, Melbourne.

Wiltshire RJ, Barclay N, Kane HE, Kenny FR, Machin B, *et al.* (1962b) Transcript of proceedings before the State Development Committee held at Melbourne on Thursday 1st February: inquiry into the introduction of European carp into Victorian waters. A. C. Brooks, Government Printer, Melbourne.

Winemiller KO, Rose KA (1992) Patterns of life-history diversification in North American fishes: implications for population regulation. *Canadian Journal of Fisheries and Aquatic Sciences* **49**(10), 2196–2218. doi:10.1139/f92-242

Wisniewski CD, Diggle JE, Patil JG (2015) Managing and eradicating carp: a Tasmanian experience. In *New Zealand Invasive Fish Management Handbook.* (Eds KJ Collier, NPJ Grainger) pp. 82–94. Lake Ecosystem Restoration, New Zealand.

Wood P, Barker S (2000) Old industrial mill ponds: a neglected ecological resource. *Applied Geography* **20**(1), 65–81. doi:10.1016/S0143-6228(99)00015-6

Woon B (2024) *A Brief History of Carp Fishing: Why is it so Popular?* [online]. Available: <https://badangling.com/carp/a-brief-history-of-carp-fishing/> (accessed 15 January 2024).

Yick JL, Wisniewski C, Diggle J, Patil JG (2021) Eradication of the invasive common carp, *Cyprinus carpio* from a large lake: lessons and insights from the Tasmanian experience. *Fishes* 6(1), 6. doi:10.3390/fishes6010006

Yokomizo H, Possingham HP, Thomas MB, Buckley YM (2009) Managing the impact of invasive species: the value of knowing the density–impact curve. *Ecological Applications* **19**(2), 376–386. doi:10.1890/08-0442.1

You X-L (2004) The earliest historical record of pisciculture in China and the author of an ancient guidebook to pisciculture in China. *Chinese Journal of Zoology Peking* **39**, 115–118.

Younger RM (1970) *Australia and the Australians*. Rigby, Adelaide.

Yuan ZH, Huang W, Liu SK, Xu P, Dunham R, *et al.* (2018) Historical demography of common carp estimated from individuals collected from various parts of the world using the pairwise sequentially markovian coalescent approach. *Genetica* **146**(2), 235–241. doi:10.1007/s10709-017-0006-7

Zeder MA (2012) Pathways to animal domestication. In *Biodiversity in Agriculture: Domestication, Evolution, and Sustainability*. (Eds P Gepts, TR Famula and RL Bettinger) pp. 227–259. Cambridge University Press, Cambridge.

Zen Nippon Airinkai (2024) *New Category of the Varieties and Sizes at ZNA International Koi Show* [online]. Available: <https://zna.jp/en/new-category-of-the-varieties-and-sizes-at-zna-international-koi-show/> (accessed 10 October 2024).

Zeuner FE (1963) *A History of Domesticated Animals*. Harper & Row, London.

Zhao Y, Gozlan RE, Zhang C (2015) Current state of freshwater fisheries in China. In *Freshwater Fisheries Ecology*. (Ed. JF Craig) pp. 221–230. John Wiley & Sons, Oxford.

Zhou B, Wu R, Randall D, Lam P, Ip Y, *et al.* (2000) Metabolic adjustments in the common carp during prolonged hypoxia. *Journal of Fish Biology* **57**(5), 1160–1171. doi:10.1111/j.1095-8649.2000.tb00478.x

Zhou J (1990) Cyprinidae du Miocène moyen du bassin de Shanwang. *Gujizhui dongwu xuebao* **28**, 95–127.

Zhou J, Wu Q, Wang Z, Ye Y (2004) Molecular phylogeny of three subspecies of common carp *Cyprinus carpio*, based on sequence analysis of cytochrome b and control region of mtDNA. *Journal of Zoological Systematics and Evolutionary Research* **42**, 266–269. doi:10.1111/j.1439-0469.2004.00266.x

Zoological and Acclimatisation Society of Victoria (1872–1878) *Proceedings of the Zoological and Acclimatisation Society of Victoria, and Report of the Annual Meeting of the Society*. F. A. Masterman, Melbourne.

Index

For Product Safety Concerns and Information please contact our EU
representative GPSR@taylorandfrancis.com
Taylor & Francis Verlag GmbH, Kaufingerstraße 24, 80331 München, Germany